Table of Contents

About the Author

E.K. Hein is a former middle-school language arts teacher from West Chester, Pennsylvania. Having taught language arts and English skills to students for 10 years, he was drawn to the notion of a thematic, interdisciplinary unit involving crime scene investigations after reading William Maples' *Dead Men Do Tell Tales*. He subsequently became an avid crime story reader, as well as an attendant of various special seminars at the Virginia Institute of Forensic Science and Medicine, which is designed to explore modern crime fighting techniques and conduct forensic analysis for lay people. He also spent a summer working in a toxicology lab under the direction of Dr. Fredric Rieders. He has presented excerpts of his crime unit curriculum to various teacher organizations across the country and published in several educational magazines. In 2002, his curriculum unit "Investigating Crime Scenes in Literature" was featured in *The Wall Street Journal* as a model for innovative, engaging curriculum. In 2004, he published his first book, *Partners in Crime: Integrating Language Arts and Forensic Science*, which has now become popular with teachers who are looking for interdisciplinary units in education. E.K. Hein now resides in Sparta, New Jersey, with his lovely wife and two sons and works for a publishing company.

1

Decomposing Dudes and Morphing Maggots

"It sounds like Rice Krispies," commented Melinda Monsternick.

"They don't like any kind of light, so when we turn the body over, they squirm away into the ground."

"You mean soup. The ground looks like a mud puddle. I've been a detective for 20 years, but I have never seen decomposition like this."

The research facility at National Medical Labs on the outskirts of Philadelphia consisted of a warehouse-sized building that served as the crime lab, and the research facility, where different outside environments were created to measure and study the decomposition rates of human bodies that had been donated in the name of science. Some bodies were buried; others were put in the trunk of a car or wrapped in a carpet and placed in a pool of water. As many as 10 bodies could be found decomposing within the facility at any given time.

Dr. Blass tried to explain the process of decomposition and its relation to forensic entomology. "It's disgusting, but bugs and other fauna are like clocks of death," he said. "We're talking about the PMI here, and I don't mean the thing that happens to women once a month. The post-mortem interval is comprised of several factors. We need to consider the amount of rigor mortis, the color of the body, which is lividity; and of course, the algor mortis, which is the temperature. But you know all of that already, don't you, Monsternick?"

"Yes, I had a training class recently on time of death and using bugs for identification."

"Right. That's been my passion for many years now. I love bugs. My wife would like to say otherwise, but I'm one lucky duck. When I first started this job, I was rearing bugs in my basement. One time, I used gauze around the top of my rearing chamber to let in some air. Bad move. They worked their way through the small openings in the gauze that I thought would never be possible. Of course, the maggots had not finished their post-feeding stage and decided to hide all over my basement. Before you knew it, flies were buzzing all over the house. My wife started swatting them with the newspaper and yelling at me. I assured her it wouldn't happen again and replaced the newspaper with a small butterfly net. We spent the next week chasing blowflies and flesh flies around the house to pin in my first bug collection. Since then, I've kept everything in the shed and away from my wife. I also had to buy a lot of jewelry and flowers to make up for my past discrepancies."

"I would imagine so," laughed Melinda. "If you're not in the business, bless your heart for having a sense of humor and love."

Dr. Blass was a world-renowned forensic entomologist who was very learned in the area of decomposition and the process of insect infestation. His research has seen bodies placed in streams, car trunks, hanging in branches, in the shade, sun, and every other place imaginable, all in the name of science. He continued explaining how decomposition works.

"The first stage is the fresh stage, or autolysis. After your heart stops beating, your body becomes anaerobic, which means there's no oxygen present. What is happening is the bacteria in your body that was once good for you is now detrimental because it breaks down your insides. I mean blood vessels, tissues, organs, and the like. It all becomes food for the bacteria."

As Dr. Blass was expounding on his grotesque studies, Melinda sat intently munching on a granola bar and sipping hot coffee. She did not mind the inner workings of the rank bacteria that consumed the body, for she had grown up in the Poconos with two brothers who had hunted and fished. It was not uncommon for them to entertain themselves at her expense. One time, they'd left a deer head on the handlebars of her bike. Eventually, her repulsion diminished and she became desensitized. Chuckling to herself, she continued her education on decomposition.

"The bloating stage begins when the stomach begins to rise," explained the good doctor. "While all of those enzymes are eating away at your insides, they release gases, most as a methane gas. You know, the stuff that farts are made of. Well, as that gas builds up, it causes the body to grow like the fat lady at the circus. Of course, if there are lots of wounds in the body, the bloating might not be apparent because the gas has a place to go. The tissue actually separates from the muscles because of gas pressure. This is also called skin slippage. This works really well if you want to slip off the skin from a hand and fingerprint it. One time, we wanted to see if we could use a bloating pig as a Bunsen burner. We cut a hole in the stomach, and a nice blue flame burned for fifteen minutes. Stinky as hell, but interesting.

"In the middle of the bloating stage is usually when most maggots come to life and start chowing. These guys have a built-in digestion system. They spit out enzymes that will start breaking down food before it is even eaten. Talk about time efficiency. Strength in numbers is their mantra, so they work together as a mass that consumes the body. As they squirm and squish together, the friction against the dead body causes the internal temperature to rise significantly. If maggots are present, you cannot rely on an accurate body temperature."

Melinda nodded, impressed with his scientific claims. She was kind of bummed because her new sidekick, Fredric Hassloch, was not with her. He had just joined the unit and was in training to become a detective, and would have found this impromptu lesson helpful. She made a note to herself to remember as much as she could and listened on eagerly.

"Next comes the putrefaction stage," claimed Dr. Blass. "Tissues begin to liquefy within the body and leak out. These fluids, combined with ammonia from the maggots, seep into the soil, and it becomes alkaline. At this point, none of the normal microscopic inhabitants are happy, so they leave and the others take over their turf. Years after the body has skeletonized, the soil can be tested with a Berlese funnel and the microorganisms will still be there.

"While these harmless little maggots are enjoying their grub, other predators come to feast on the maggots. They might be wasps, ants, bees, and beetles. I have seen a string of ants marching with maggots. It looks like a little white dotted line; truly amazing. I guess they have nutritional value or else the bugs wouldn't eat them. Other times, the hairy maggot will come and attack other species of maggots. They open the maggot and suck out their insides like Dracula.

"As we near the end of the process, we hit the decay stage. The skin is so badly decomposed, it tears and breaks, allowing the liquid insides to ooze out. You can easily see the maggot masses ravishing the flesh as the skin breaks apart. Maggot masses are most evident because they have found their munching buddies and are working hard. You need a strong stomach for this stage because it smells the worst. Think of rotten eggs, beef, and milk all sitting in a bag in a hundred-degree sun. Open the bag, and you're pretty close to the smell of decay. Near the end of this stage, some species of beetles will arrive to gnaw the dry skin.

"After the gorging, the maggots leave for a post-feeding stage. They will find safe ground and begin to pupate. This is the final larval stage and when we see them next, they have metamorphosed into adult flies. By the end of the decay stage, only 20 percent of the corpse remains, consisting of skin and bone. When the corpse is reduced to skin, bone, and cartilage, the flies leave the body. The corpse is now beginning the post-decay stage. You will find various species of beetles, like hide and hister, that eat the skin. The soil is also being overrun with microorganisms that continue to feed on decomposing material.

"Now the only thing left is the skeleton. Believe it or not, hunters and fishermen have discovered many bodies in the woods. Do you know why?"

"Yes, I do. My brothers found a body one time while hunting in the woods. Before they decided to tell anyone about it, they made me go with them to check it out. It was a half-exposed human skeleton. We sat there poking the skull with a stick and a few bugs ran out of it. Right then and there I was hooked. I was instantly wondering who the person was, and it made me sad because someone had died. That became my impetus for wanting to become a detective. I wanted to fight crime and keep people from getting killed. I also thought about studying one of the sciences, but decided I didn't want to work in a lab."

"It's amazing how this stuff works like clockwork, detective. I am trying to develop an instrument that can electronically sniff the air and tell me the exact prevailing compounds. It might assist me in determining the PMI within minutes."

"So what about the maggots?" inquired Melinda "I'd like to know a little more about them. I brought many samples over for you this morning. I packed them in dry ice, too, just like you requested. I hear you can actually get DNA from the crop of the maggot."

"Yes, you can," replied Dr. Blass. "The research is still emerging, but the crop, otherwise known as the belly, can glean much information about the body on which it feasts. For instance, we had a body where the person was on cocaine, and the maggots worked faster because of the drug. It works on the converse, too, with other drugs. Let me tell you a little about our friendly rice cakes down there.

"After the female fly eats her share, she produces eggs and usually lays them on an open orifice or area that has been traumatized. This way, when her babies are born, they have instant access to sustenance.

"Maggots fatten up as they continue eating. The weather has a lot to do with how hungry they are. Through our research, we have acquired an exorbitant amount of data on maggot growth. All maggots will shed their skin three times. The outside of the maggot is composed of chitin, which provides a safety net from the environment. If you think about your nails, they are made of chitin as well. They will not shed into the next instar stage until they can fill the cuticle.

"In the first instar stage, the maggot is the size of a grain of rice. In scientific terms, that equates to approximately 5 millimeters. They will remain in this stage anywhere from a half a day to one and a half days. Remember, it all depends on weather conditions.

"The second stage of the instar can last anywhere from less than half a day to over three days. At this point, they are hanging with their homies, so to speak, as they dine on their smorgasbord. This is also when the body temperature can reach higher temperatures, like I said before. You will also see maggots with two dots at one end. This is where they breathe. In the scientific world and court, we call them posterior spiracles, but here at the facility, we call them butt breathers. The maggot has also doubled in size and will go almost a day before reaching the third and final stage."

"You mean they breathe through their butt?" inquired Melinda.

"Yes, they do," answered Dr. Blass. "If you had your face buried in a slab of beef, you would have to breathe through your butt, too, for continual nourishment."

She agreed, thinking about when her brothers would pinch her nose shut when she was sleeping. Bizarre entertainment, but for teenage boys, it was customary. Dr. Blass picked up one of the maggots on the end of his pen knife.

"This guy here is in his last stage before entering the post-feeding portion of his life. As you can see, he has tripled in size, and you really have a nice view of the crop. The ridges of the cuticle are easily seen and he now has four spiracles to breathe. This might last anywhere from one to five days. After his pie hole is stuffed to the hilt, he wanders off to find a cozy resting place to digest and get some rest and relaxation."

"Doc, how many bodies have you studied over the years? I find these averages of fly development staggering."

"That's putting it lightly, kiddo," said Dr. Blass. "The body we are looking at here is guest number 832 at the facility. The data we have collected over the years is insurmountable. God bless the computer. Anyway, let me finish my enlightening lecture.

"Maggots seek shelter away from distractions on the ground surface, so they usually go under rocks, leaves, or logs to transform. It pupates to a tiny brown oblong object, like the pigskin used on Sunday afternoons in the fall. The inside of the maggot goes through a metamorphosis where it turns into a fly. When it pops out of its casing, it is very light in color, almost beige. Within a few hours, the fly gets a tan and turns into its normal color depending on the species.

"The whole process ranges from 10 days to a month. The cycle repeats itself until there is no more flesh to eat. Then it has to seek out other manna for nourishment. Long live the fly!

"We did an experiment once to see how far away the flies could smell death. We had some blowflies that we marked with some orange paint. After we walked approximately a mile from the farm, we released the flies and laid out a fresh corpse that was donated to us by a lovely gentleman who wished to continue his name in science. Within two hours of the release, they were at the body eating and laying eggs. The other flies were a little apprehensive about the orange dots on them, but everyone got along for once. It was a great experiment. I published that one in the *Journal of Forensic Science*. I could go on for hours, but I bet you have other things to do, Detective."

"Yes, I do, but I love listening to this. I'm hoping your research will help us break this case. The body we found is still at the morgue. There was an odious smell infiltrating the woods when we arrived. We are waiting for the DNA results to come back to get a possible ID. We're hoping it might match the missing kid we've been looking for. Once we get further into our investigation, I'll be in touch. If we can show the time of death within a few hours, maybe we can find a perp without an alibi for that time period."

Melinda Monsternick finished her morning coffee, thanked the doctor, and went on her way. She had to pick up some cat litter for big Will and then head over to Fredric's house for his lesson of the day; you guessed it, decomposition. Since he was the new guy she called him Hassie, which was short for his last name, Hassloch. She planned on giving him some homework and sending him to the library to do research to help their case.

∾ ∽

Vocabulary

Autolysis   Also called the fresh stage, the beginning stage of decay, marked by the breakdown of blood and tissue inside the body.

Bloating   The second stage in decomposition, marked by the release of gases in the body, which causes it to rise and appear bloated.

Decay and putrefaction   The third stage in the decomposition process, marked by changes in the outward appearance of the body. It is also the stinkiest because of the release of gases.

DNA   Deoxyribonucleic acid.

Entomology   The study of insects, their life cycles, and their habits.

Instar   One of the three stages a maggot goes through during development.

Lividity   The changing color of the outward appearance of the body as it enters different stages of decomposition.

Maggot   Newly hatched larvae of an insect.

PMI (post-mortem interval)   The amount of time that has passed since the death and the discovery of the body.

Rigor mortis   After death, the stiffening of joints, muscles, and tissues in the body. It can last anywhere from 12 to 36 hours.

Skeletal stage   The final stage of decomposition, marked by bone that has been stripped of tissue, muscle, and organs.

Background Information

Although gross, bugs can tell us a lot about the time of death. Entomologists are becoming more valuable in the forensic world. Research is underway to acquire DNA samples from the crop or belly of a maggot. Insects and fauna represent different stages of decomposition and therefore can help entomologists determine the approximate time of death.

PMI (post-mortem interval) is determined through body temperature, lividity, and rigor mortis. Lividity comes from the Latin word meaning "color of death." The body changes color as decomposition progresses.

There are other factors affecting PMI as well, such as ambient air temperature, drugs in the body, or different environments (water, wrapped in something, buried, etc.). Rigor mortis comes from the Latin term for "stiffness of death." When the body stops pumping blood, proteins known as actin and myosin form strong bridges that cause the muscles to stiffen. Rigor mortis usually lasts from 12 to 36 hours. This also helps determine the approximate time of death. Another strong indicator of PMI is the type of insects present on and around the corpse. These bugs can be collected, reared, and identified.

Stages of Decomposition

The first stage is the *fresh stage* or *autolysis*. It starts at the moment of death and ends when bloating is evident. The corpse does not change much on its outward appearance. This is when the inside of the body begins its changes. Tissues break down along with blood vessels. Bacteria begin to spread through the body, releasing enzymes that consume the tissue.

Next comes the *bloating stage*, which begins when the stomach begins to rise. At this point, putrefaction becomes apparent. The enzymes break down and digest tissue and organs within the body. The by-product of the metabolic activity of the enzymes release gases in the form of methane. As the methane increases, the body begins to swell like a balloon. The methane, along with other gases, creates the sweet-sickly smell of death that is found around decomposing material.

During the bloated stage, fluids seep from the body. These fluids, combined with the ammonia from the maggots, seep into the soil, and it becomes alkaline. This is when the normal inhabitants of the soil move out and the microscopic mites move in to feed. They can remain there for years after the body has decayed.

The third stage of *decay and putrefaction* is also marked by a physical change. This begins when the skin is broken and gases are released from the body. Large masses of maggots are highly noticeable in the early to middle stage of decay. The corpse is still moist and large amounts of fluids are seeping into the soil. This is also when the corpse emits the foulest odor. In the latter portion of this stage, the amount of beetle activity increases. At this point, the maggots leave for a post-feeding stage. They will find safe ground and begin to pupate into their final larval stage. When seen next, they will have metamorphosed into adult flies. By the end of the decay stage, only 20 percent of the corpse remains, consisting of skin and bone. When the corpse is reduced to skin, bone, and cartilage, the flies leave the body.

The corpse now begins the *post-decay stage*. This is when hide beetles come in to feed on the skin. Rove and hister beetles accompany the hide beetle to feed, along with other insects.

If the area is wet, the beetles may not appear. They need dried skin to eat. Maggots will also stay on the corpse if it is wet. By the end of the post-decay stage, only 10 percent of the corpse is left.

The final stage is the *skeletal stage*. The corpse is reduced to bones and hair. If left out in the sun, the bones will be bleached white. The only activity left is in the soil. These microscopic organisms can be extracted using a Berlese funnel. A sample is placed in a funnel and a 60-watt or higher bulb is placed over it. A beaker with preservation liquid is placed in a beaker at the bottom of the funnel. As the microorganisms migrate to the bottom of the funnel and away from the light, they fall out into the solution, where they can be viewed under a microscope and identified.

Decomposition is dependent on many factors related to the environment. Bodies have been preserved for thousands of years, such as the Iceman found in Europe. However, in the right conditions, a 150-pound body can be reduced to a skeleton in two weeks.

Fly Development

After the fly has eaten her share of flesh, she will oviposit her eggs on a corpse. Most likely, it will be in an area that is an open orifice or an area that has exposed flesh due to some sort of injury or trauma.

Laying eggs.   **Maggot mass.**

The greatest numbers of fly eggs occur in the early to middle portions of the bloating stage. The maggots release an enzyme that predigests the food, allowing for easy ingestion With the enzymes breaking down the tissue, a semiliquid is created before it is consumed by the maggot. The maggots will work together and create a mass that moves throughout the body and ravishes it.

The metabolic changes from the maggots and the anaerobic bacteria may cause the body's internal temperature to rise. It has been recorded as high as 127 degrees.

As the maggots grow, other insects are attracted to the corpse. Bees, wasps, hornets, ants, beetles, and other types of flies come to feed on the maggots and eggs.

Maggots develop in size as they eat. Their eating cycle is dependent on the ambient air temperature and weather conditions. They will go through three instar stages. The external cuticle is composed of chitin. It is flexible and provides protection from the environment. It does limit their size, so it cannot molt to the next instar stage until it grows large enough to fill the cuticle.

The first instar usually takes the least amount of time. Most maggots complete this stage in 11 to 38 hours, with most developing between 22 to 28 hours. At this

First instar.

Second instar.

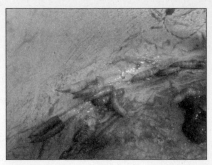

Third instar.

time, the maggot is approximately 5 millimeters in length—about the size of a grain of rice.

The second instar stage runs from 8 to 54 hours. Maggots develop a maggot "mass" that continually devours flesh. Maggots need to breathe while eating, so they develop posterior or anal spiracles. Two posterior spiracles are present, and at this stage they are approximately 10 millimeters in length. Most species will go 11 to 22 hours before molting into the third stage.

The third stage lasts the longest. It is divided into two parts. In the first part, there is continual feeding on the corpse. This can last anywhere from 20 to 96 hours. Maggots have four developed posterior spiracles, and at this point, are approximately 17 millimeters in length. When it cannot eat anymore, it enters the second part, which is the post-feeding or wandering stage. This stage can last from 40 to 504 hours, but the average time is 80 to 112 hours. During this time, the stomach of the maggot begins to empty and it starts the pupating stage.

Maggots move to a drier area where they will not be disturbed to finish the metamorphosis. When they do begin to pupate, the color of the pupa is white to yellow. Over the next few hours, the maggot will turn deep reddish brown. At this point, the casing looks like a football.

It is resistant to heat, cold, and flooding. While in the case, the larva changes into an adult fly. The tissues and structures of the maggot are dissolved by a process called histolysis. New structures begin to develop such as legs, eyes, and wings. This stage can take 4 to 18 days, but most will last 4 to 16 days.

Now the fly is ready to emerge through a round seam at the top of the case that the fly will pop off. The new fly does not look like an adult fly because it is light in color and soft while the wings are wrinkled and collapsed. Over a period of hours, the cuticle hardens and assumes its normal color. Fly development from beginning to end can take from 10 to 27 days at a temperature of 80 degrees. Their adult lifespan will last from 17 to 39 days, and they can reproduce 5 to 18 days after emerging.

Forensic Implications and Jurisprudence

When maggots or other insects are present at a crime scene, many samples are taken. Some are sent to the lab to be reared, while others are preserved to show the developmental stage at the time of discovery. The insects that are kept alive are raised to adulthood and allowed to reproduce. They are placed in vials with small pieces of beef liver for sustenance. When the maggots in the lab reach the same stage as the ones found at the crime scene, they are compared for time estimations. In order for successful results, the scientist should simulate exact conditions in which the samples were found. Since the surrounding climate is one of the most important factors to growth development, it is vital these conditions be recreated.

After identification has been made, the entomologist will try to determine what kind of species of insect it is. Blowflies and houseflies are usually the first ones to arrive, while flesh flies come later when the flesh is exposed, for the flesh fly lays live maggots instead of eggs. There are a considerable variety of flies on this earth, and only a specialist can make the identification.

The court will determine the qualification of an expert and has the right to accept or dismiss a person as an "expert." While in court, the entomologist can testify only about information dealing with insects, the rearing process, and the results of that particular experiment. If pictures are required to elaborate on the process of identification, it may be done with permission of the court. Most pictures of decomposition and insect infestation are horrific, if not traumatizing to the victim's family, so the court must be considerate in their decision.

A qualified entomologist must possess a degree in one of the sciences with extensive training and background in entomology. It is also recommended that he or she participate in numerous experiments and have published material on the subject.

2

Spurts, Spatters, and Stains

The alarm rang and Fredric Hassloch rolled over in his bed. He was exhausted after spending the night diligently filling out reports on a decomposed body that was discovered in Fairmount Park. He was slowly learning the process of crime scene integrity and how to complete reports correctly to avoid any kind of debacle with defense attorneys. They were merciless when asking ridiculous question after question trying to find a loophole to get their clients off. Fredric had not yet been privy to the courtroom shredding by pedantic lawyers with one thing on their mind: exploiting their intelligence and wits.

Fredric hit the snooze button a few more times until he heard his doorbell ring. He drowsily shuffled to the door and opened it, where Melinda Monsternick stood with two coffees and some bagels in hand. "Good morning, sunshine," she said with a perky smile. "I trust you've recuperated from the interminable paperwork and are ready for a fresh start. I had an interesting time over at National Medical Labs this morning with Dr. Blass. He told me all about decomposition and how flies lay their eggs in open orifices for baby nourishment and the leaking of fluids when putrefaction begins. Want a bagel and some coffee?"

Fredric squinted his eyes as if the light was an evil invader of his vision. "What are you talking about? It's my day off, Monsternick. I was thinking of running some errands today."

"Well, you'll have to put them on hold. We got another body over on Ridge Pike. Young man with a fatal gunshot wound to the head. Looks like he was shot execution-style. Might be a connection with the body we found yesterday who had a similar gunshot wound to the head. So, get dressed, get some coffee pumping through the veins, and let's have a lesson in DNA collection."

Fredric was still bewildered. His head felt a little foggy as he tried to internalize the new information. It started to come together in his head as he recalled the paperwork from the night before. "Do I have time to go to the bathroom and brush my teeth?" he asked with a yawn.

"I guess, but we need to go. The EMTs are waiting for us so they can take the body. You have two minutes. Go!"

Three minutes later, Fredric and Melinda were on the Schukyll Expressway en route to another possible murder scene. It was plausible that there was a potential serial killer running around Philadelphia, but coincidences happen. Over the years, Melinda had seen execution-style murders with Mafia- and gang-related shootings. Those killers who had been prosecuted had claimed a sense of domination over their victims. She thought it was more of a male ego kind of thing that she would never understand, being a woman who would rather exact her revenge differently.

When they arrived at the crime scene, tape had been placed around the front porch and an officer was standing there, clipboard in hand. "Hey, guys," he said. "You need to sign here, and Hassie, I need your fingerprints when you're done. I don't have you in the system yet."

"Don't worry, I put them in AFIS the other day," replied Fredric.

"Oh, cool. Do you know what that stands for, rookie man?" the officer joked.

"Automated Fingerprint Identification System. Come on, that's easy."

Fredric was beginning to comprehend the legal process of approaching and documenting crime scenes. Of the few crime scenes he'd attended,

Melinda was responsible for his formal education as a crime scene investigator. It was customary for anyone who entered the crime scene to fill out the log and provide fingerprints, so they were not mistaken for a possible perp. One time, Melinda was investigating a robbery dealing with a pillaged jewelry box. After she dusted it and prints appeared, she took it to County to run through the system. Unfortunately, she took off her gloves after dusting the box, so when the technician went to dust the box a second time, her prints showed up, and much to her chagrin, she was busted. Therefore, it was paramount that the only people who entered the crime scene were relevant to the immediate investigation. Therefore, after a path was set up through the house, detectives, EMTs, and the coroner were the only ones permitted to enter.

"All right, Hassie," said Melinda. "Let's take a look here. What are your first thoughts?"

"The place is a mess, probably from a fight or some kind of struggle. There's blood spatter on the wall and some more blood on the hardwood floors in the living room." As Fredric walked into the kitchen more bloodstains were apparent. "It looks like there was some profuse bleeding in here. Look at these long stains on the wall."

"Those are consistent with an arterial spurt," replied Melinda. "It happens when someone has a major vein slashed. From the looks of the guy on the floor, it could be either the jugular or carotid. Get the camera so we can take some shots. I also want video. Remember to use the mini-rulers so we have a reference for size. You don't want to suffer the wrath of a hungry defense attorney."

"Right," said Fredric. "And I'll make sure to shoot at a 90-degree angle for the best results."

The two documented all possible pieces of evidence and filmed every room. The microphone on the video camera was turned off because background noise and people talking somehow made the film less effective. At this point in the investigation, the EMTs came to take away the body. It would go to the morgue for an autopsy. "It's kind of odd with the execution-style gunshot," noted Melinda. "If he had a major vein slashed, he would have bled to death. The gunshot seems a little excessive. Maybe the perp knew his victim. I had a case once where a girl was stabbed 47 times. It turns out she was cheating on her boyfriend and he was so pissed, he said he couldn't stop himself. It's amazing what happens in the heat of the moment. Hassie, the longer you work this job the more you realize the insanity of our fellow humans."

Evidence collection was the next step in the process. "The first thing we have to do is run a presumptive test to make sure it's blood. Take a swab and wet it with some distilled water, then put two drops of KM reagent on it. It if turns pink, you can presume it's blood."

"KM who?" Fredric asked.

"Kastle-Meyer test. It's basically phenolphthalein. It reacts with the hemoglobins in the blood to cause a color change. After it tests positive we move on to step two. We will need swabs in the general areas of the bloodstains themselves. Hopefully, Arthur will be able to obtain more than one profile. Hassie, I know you haven't done this before, so let me explain. First, all samples must dry before we package them. I say 'package,' because we never use plastic when collecting possible DNA evidence. If you place an item in a closed bag with no air, condensation will occur, promote bacteria growth, and possibly destroy the DNA. The environment is our worst enemy. People who think they're smart dump bodies in rivers, lakes, and the ocean to wash away evidence. Yes, it washes away some evidence, but if there is any trauma to the body, it will still be evident through X-rays or autopsy.

"We need to double up on gloves so we're always covered, and we change the outer pair after each collection," Melinda continued. "From looking at these stains, they might have been here awhile. Hassie, blood dries from the outside in. These arterial stains are pretty dry, but the puddles on the floor are still wet. Let's collect the wet ones first. Take one of the single-sided swabs out of the kit. Then hold the tip perpendicular to the stain and try to concentrate as much blood as you can on the tip. This makes it easier to extract a DNA sample at the lab. But before you do that, fill out one of the evidence forms. Make sure you put your badge number, date, time, where you found the evidence, condition of the evidence, and how you handled it. The swab can dry over in the drying container we brought. I'll go over and take some samples from the arterial spurt."

"Is that the same collection process?" asked Fredric.

"No, I'll have to wet the tip with distilled water. Then I have to ever-so-gently rub the stain to concentrate it at the head of the swab. Then it can dry and be packaged."

"What if the stains are too dry to rub off?" fired back Fredric.

"Then we take a clean sterilized scalpel and scrape it into a paper pill box or envelope. Same information, too."

After the DNA evidence was collected, the dynamic duo searched tediously for other sources of evidence. In the kitchen sink, Fredric found a Henkels serrated bread knife. It was covered in blood, which was also smeared on the left side of the sink. He photographed it and placed the knife in a paper box with the pertinent information. As he scanned the kitchen, he noticed a single bullet casing shining like a golden light offering itself from heaven. It was almost as if the casing was standing out because it wanted to be found. Fredric's dogmatic side was not as strong as his sense of being in the right place at the right time. He picked up the casing with his pen cap and started thinking about the body on its way to the morgue. Furrowing his brows, he tried to recreate the crime in his head. Where did it start or end? Questions poured into his brain as he pondered the possibilities.

His thoughts were interrupted by the screaming sobs of a woman who was standing in the doorway. It was the victim's mother. The officers were trying to take her outside, but she squirmed out of their grasp. She was looking at the blood-stained house as she fell to the floor crying. Melinda immediately ran over to her and took her into another room.

"Ma'am, I'm so sorry for your loss. You shouldn't be here. This place is a mess and I don't want you to see your house like this."

"My son! What happened to him? Where is his body? Who did this to him? Oh my God, this is horrible and I can't stand it. Help me, please!" Her words were barely understandable, for she was beginning to hyperventilate between her sobs.

Melinda brought her back outside and signaled one of the officers to take her somewhere and try to calm her down. She was hysterical at this point and a sedative would be something to consider. Melinda returned to the house with a heavy heart. She could not comprehend the thought of losing a son. She had no children except for her cat, big Will. He was like her child and she would be very upset if anything happened to him, but it was obviously nowhere near what this lady was going through.

Adrian Belkoy was the second body found with an execution-style gunshot wound to the head in the last two days. Melinda immediately thought of Phillipe Mindrago, an 18-year-old student at Roxborough High School, who had been missing for four weeks. She thought about the decomposing body they found in Fairmount Park yesterday. Could that be him? Both boys were seniors and prominent athletes within the school. Belkoy had been an All-American in football, and Mindrago was a budding basketball star who already had a full paid scholarship to Duke the upcoming fall. Both were honor roll students and members of student government. Was it a matter of coincidence or were these murders premeditated with fatal intentions? Fredric was prudently filling out his evidence bag before placing the bullet casing inside when Melinda interrupted his thoughts with the clacking of her heels on the kitchen floor. "What's the word, Hassie?" she asked.

"Well, I have a possible weapon here covered in blood, and a bullet casing on the floor. Too bad we didn't find one yesterday with the other body. What do you think is going on here?"

"I would say a possible coincidence, but these two are approximately one month apart in time of death. Let's assume for a moment that the other body is Phillipe Mindrago. It would be more than coincidence because they both go to the same school, have similar goals and aspirations, and were killed in the same manner. Call me a Celestine Prophecy debunker, but I think there is a connection that is not fate. Who would want these two dead? They seem to be poised denizens of the city of Brotherly Love. What a shame. We'll have to make our rounds at their school and see what we can find out. You know, talk to teachers, administrators, kids, coaches, and parents. Obviously, someone had a score to settle and wanted to do it with authority, hence the shots to the back of the head. Hassie, we could be looking at jealous teammates or schoolmates, or some sad sap who was bullied by these two. Let's hope Mindrago really is missing and he'll turn up soon. For now, let's be as objective as possible. We don't have a solid starting point, so we'll have to wait until we go to the school tomorrow. For now, we have to finish up here and head over to National Medical Labs to drop off this evidence. Dr. DiGregorio can give us a better idea of what we're looking at once we have the autopsy results."

With that, they finished their job and secured the area until further notice. As they walked out the door, the smell of spring was in the air. The snow had melted and small buds were beginning to appear on the tree limbs as the forsythia started peeking through in its yellow splendor.

⌁   ⌁

Vocabulary

AFIS   Automated Fingerprint Identification System. A computer system that electronically compares unknown latent prints to prints in a database of known persons. The computer produces a list of possible matches that meet the satisfying criteria. An official print examiner then determines whether or not there is a definitive match.

Arterial spray   Bloodstain pattern caused when a major artery or vein is cut open.

Ballistics   The study of projectiles and the direction of travel.

Blood spatter   The pattern that is created when blood comes in contact with a surface or is acted upon by an outside force.

Cast off   The pattern that is created when an instrument covered with blood moves through space at medium- to high-speed velocities.

Contact pattern   The pattern that is created when a bloody object or body part comes in contact with a surface.

Crowning   Little splashes that appear outside of the circle of a blood drop.

Hemoglobin   An iron-containing protein in red blood cells that combines reversibly with oxygen and transports it from the lungs to body tissues.

High impact   The pattern that is created when blood is acted upon by an outside force at high velocity.

Impact angle   The angle in which blood strikes a surface.

KM test (Kastle-Meyer test)   A presumptive test for all kinds of blood.

Luminol   A reagent that reacts with the hemoglobins in the blood and makes it glow.

Passive drop   A pattern that is created when blood traveling under normal conditions comes in contact with a surface.

PCR (Polymerase Chain Reaction)   A DNA amplification technique.

Smear   A pattern that appears when a bloody object or body part comes in contact with a surface.

Surface tension   The elasticlike property of a liquid's surface that makes it tend to contract, caused by the forces of attraction between the molecules.

Background Information

Bloodstain evidence is crucial at a crime scene. A valuable source of DNA, it also helps investigators determine the position of people or objects used in the crime as well as track the movement of a person and show the sequence of events that may have taken place. Bloodstain evidence, referred to as blood "spatter," means that blood has predictable patterns of flow when acted upon by gravity, an outside force, or any given surface that comes in contact with the blood.

Some criminals attempt to cover up or remove bloodstains by washing them with bleach or other heavy-duty cleaning products. Unfortunately for the bad guy, Luminol reacts with the hemoglobin in the blood, and the smeared and previously cleaned stains will light up the room like a fluorescent light show. In the past, Luminol has not been the first choice of the crime scene investigator because it degrades the DNA sample and might not show up under the use of RFLP (Restricted

Fragment Length Polymorphism). However, with the advent of PCR (Polymerase Chain Reaction) techniques, degraded samples in small amounts will most likely yield a DNA profile.

Blood spatter analysis provides two valuable pieces of information: direction in which the blood travels, and the impact angle at which the blood strikes a surface. Therefore, mathematics and physics are heavily involved in blood analysis. Surface tension is another major factor because it determines the pattern of the bloodstain. Surface tension is the elasticlike property of the liquid's surface that makes it tend to contract, caused by the forces of attraction between the molecules. Cohesive forces tend to resist penetration and separation, so in order for a blood drop to fall, the gravitational force acting on the blood

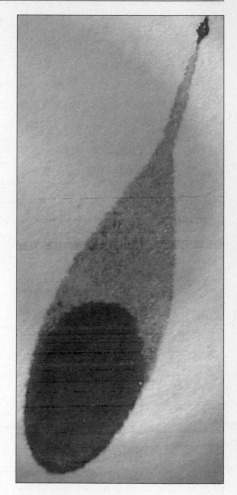

Blood drop pointing up.

must exceed its surface tension, otherwise the blood would act like a slime or goo. Surface tension also affects the impact on a smooth, hard surface by minimizing the spatter.

The direction of a blood drop can be easily observed by identifying the tail of the blood drop. Any drop of blood that strikes a surface at an angle less than 90 degrees usually creates a tear shape or become elongated into an elliptical pattern. When a blood drop with a tail is present, the tail points in the direction of travel. When a tail is not present, the area where the blood appears to be thicker indicates where the blood has landed first, hence the area with less blood is the direction of travel. Be careful, because blood dries from the outside in. If no tail is present, chances are it is a passive drop, which simply means no force or velocity has been applied to it. These occur when blood drips from any height from someone or something standing still.

The impact angle is calculated by measuring the length and width of the drop. The angle is the acute angle formed between the direction of the blood drop and the plane of the surface where it lands. The impact angle is determined by measuring the length and width of the drop. For this measurement, a SIN formula must be used. The true length of the drop is measured from the beginning of the drop to the beginning of the tail. The tail is not considered part of the true length, because the force or travel has caused it to continue to lie on the surface. Use the following formula:

$$SIN \leq \frac{Width}{Length}$$

Let's say, for example, a blood drop has a width of 1.5cm and the length is 3cm. The formula will look like this:

$$SIN \leq \frac{1.5cm}{3cm}$$

$$.5 = SIN <$$
$$\leq 30 \text{ degrees}$$

Finding the point of origin and convergence of a bloodstain can be accomplished by creating a two-dimensional point for a group of stains. A line must be drawn through the long axis of each blood drop in a group of stains in an 8-by-11-inch area. As more lines are drawn, the point of intersection becomes apparent, and that is the point of convergence.

A drop of blood traveling untouched through the air takes on a spherical shape. If the surface is flat with a somewhat smooth area, a circular pattern appears. As height increases, small splashes occur around the edges. This is called the "crowning effect" because the shape looks like a crown. When the surface is at an angle, the pattern changes.

The pattern changes when an outside force interrupts or acts upon the blood drop. One of the patterns created is a high-impact spatter. This

Blood drop from 36 inches.

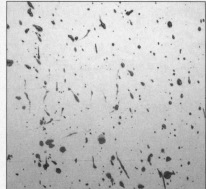

High-impact spatter.

happens when an instrument or object is used to strike a person. The high velocity of the object and impact disperses the blood and forms a pattern that is similar to a spray, with little dots or splashes of different shapes and sizes.

Another pattern is called cast-off spatter. This occurs when an object is covered with blood. Since blood easily collects on most objects, it remains there until enough force is used to pull it from the surface. This usually happens when the object is lifted or pulled back for another blow. It is done with such force that it causes the blood to leave the object and travel to the nearest surface. Because of the speed of travel, the blood has a pattern of a concentrated area with many tails casting off the stain, hence its name. These patterns are commonly found on ceilings or walls, as the object is most likely to be raised above the head.

Arterial spray or spurt is another type of pattern that obviously results from the cutting or slashing of a major vein or artery. Since the bloodstream is under pressure, any break in the flow causes a spurt. Imagine cutting a garden hose while the water is on.

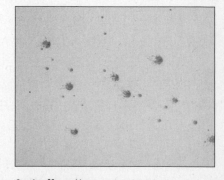

Cast-off spatter.

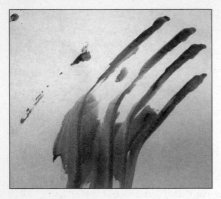

Hand smear.

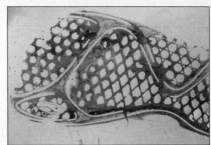

Bloody footprint.

The water continues to spray or spurt out until the water is turned off. Eventually the water runs out and the spray becomes less intense. The same premise is true for an artery or vein. Because of the pressure, the initial spurt can travel upward of 6 to 10 feet.

If a gun is used at very close range, a back spatter pattern occurs. When the projectile impacts the body with force, it forces blood and parts of flesh back in the direction of the shooter. These stains are sometimes seen on the clothes of a shooter or even in the barrel of the gun, which can be seen under a stereomicroscope.

Contact smears and transfer patterns may also be found at a scene. They are created by hands, feet, or a bloodstained object or piece of clothing that rubs against a surface. They may also show the direction or movement of a person throughout an area.

Collection Methods

In the field, there are various presumptive tests for blood. A presumptive test does not confirm that a sample is blood, but presumes it has the potential to be blood. Also, it does not determine whether it is human or animal blood. One test that is used is called the KM (Kastle-Meyer) test. It combines phenolphthalein with potassium hydroxide and ethanol. A small amount of zinc dust is then added to prevent the solution from turning a slight pinkish color, yielding incorrect results.

When blood is present in this presumptive test, the tip of the swab will turn a pinkish color. Again it does not confirm whether it is human or animal blood. In order to do that, a precipitin test may be used. In this experiment, protein from a chicken egg is injected into a rabbit. When the serum is mixed with the egg white, the egg proteins will separate from the liquid into a cloudy substance known as precipitin. For testing purposes, the sample is placed in a gel on a glass slide next to a sample of the serum. An electric current is then passed through the glass, causing the molecules to either come together or separate in the gel. A positive result is apparent when and if the line comes together where the two meet. This line is called a precipitin line.

When bloodstains are present at a crime scene, they should be documented with photos and reference points listed in the report. It is imperative that precautions be taken when handling bloodstain evidence. Diseases such as HIV, hepatitis, and STDs could exist and infect the collector. Although one may not be infected unless there is the most minute of cuts on the skin of the hand, wearing gloves is still imperative. A double set of gloves should be used when collecting DNA evidence to prevent contamination. The collector should also wear any other necessary protective gear such as bootie shoes, a face shield, or hair covering to maintain the integrity of the evidence.

Bloodstain evidence that is considered to be fragile should be collected first because environmental factors like light and precipitation could affect it. The protocol for investigators is clear-cut and the rules must be followed at all times. All evidence is collected and packed separately. Each piece of evidence should be labeled with the name of the technician, date, time, ID number, where it was found, and how it was collected.

When articles of clothing are found, they should never be folded over due to the fact that two different samples may be present. A piece of paper can be used to separate the two stains.

All evidence containing blood is packaged in paper. If stored in plastic for too long, it can cause bacteria to grow and degrade or ruin the sample. Therefore, plastic bags should only be used if an item is soaked with blood and needs to be removed. In this case, it should be taken to a secure area and handled properly. When collecting blood evidence that is wet, if possible, submit the entire item. If it is too large, use a single-sided cotton swab to take a blood sample. Document where the blood was found and how wet it was, then let the swab air-dry before packaging. When smaller stains are found, concentrate the sample at the end of the swab by using a tight swirling motion.

For dry bloodstains, a different approach is necessary. Submit the entire item if possible, if not, use a clean fresh razor blade or scalpel to cut out the stain. Label and package it as described previously. A control sample must also be taken. This should be done with another new blade or scalpel to prevent contamination.

Forensic Implications and Jurisprudence

First and foremost, extreme caution must be taken when collecting bloodstain evidence from a crime scene. Document, document, and finally, document. If all samples have been collected and labeled properly, a potential DNA profile can be extrapolated.

Experts can be called into a courtroom to explain the stains and patterns and how they are used to piece together the events at a crime scene. Due to its predictable nature, bloodstains provide solid evidence to convict or set free an innocent person. The expert should be thoroughly trained in biology with a substantial degree of on-the-job forensic training. He or she should be ready to testify only on information related to bloodstain patterns and the analysis of those patterns. Experimentation and publishing are a plus when qualifying as an expert. The court reserves the right to accept or deny the presence of a so-called expert based on his or her credentials. Again, because of the macabre suggestions with pictures of messy crime scenes covered with blood, photos of bloodstain evidence are left to the discretion of the court system.

3

A Club for Corrupt Kids

The entrance to Roxborough High was filled with school buses as students arrived for another day of brain frying with useless information never to be used again. At least that was the thought of several of the boys who were standing along the road sucking down their last cigarette of the morning. The next respite for smoking would be lunch. Melinda and Fredric decided to sit in the car for a while and watch the students filter into the school. It would also give them time to finish the hot coffee they just acquired at Starbucks. "I remember my days of going to school in the Poconos," Melinda began. "Half of the cars were pickup trucks shelved with gun racks and a Southern rebel flag on the front. Talk about real rednecks. If your jeans didn't have a hole in the back pocket from your chew, you just weren't cool. Gosh, my seventh-grade teacher chewed tobacco during science class."

Fredric frowned in disgust. "I tried that stuff once and I puked my brains out. It was right before one of my Little League games. One of the kids told me all the baseball players do it. Ten seconds later, the chew fell out of my mouth and shreds of tobacco were stuck in my teeth. Blah!"

"It's something you have to get used to. My brothers did it all the time. I never had the inclination. Hey, look at this kid over here."

A young girl walked by sporting a two-foot-high mohawk dyed green. Underneath her black eye makeup, you could see a pretty girl who wanted to make her point: *Hey, look at me, I'm different, like a freaky punk who wants to go to med school.* As she passed, a Subaru wagon painted black and white like a cow pulled up and three guys wearing Tony Hawk T-shirts and long baggy pants got out. The windows were covered with stickers and they were lucky Melinda did not cite them for window obstruction. The crowd started to dwindle as the parents dropped off the latecomers who probably would be handed a late pass and possible detention for their few extra minutes of sleep.

Melinda and Fredric strolled into the front office, where they were greeted by one of the secretaries, a well-dressed woman who smelled of fancy French perfume. "May I help you?" she squeaked from behind her glasses.

"Yes, my name is Detective Melinda Monsternick and this is my sidekick, Fredric Hassloch. We're from the Philadelphia Police Department. May we speak to the principal, please?"

"Sure, if you'll have a seat," she replied, pointing to two empty seats in the corner.

Students were still shuffling in as parents dropped off lunches and forgotten homework assignments. A tall, skinny, bald-headed man came in to greet the two detectives. "Hello, my name is Principal Theo Monk. No relation to the famous Thelonius Monk. Come into my office so we can speak candidly. Would you like any coffee or tea?"

"No thanks, Principal Monk," answered Melinda. "We just finished some high test in the parking lot. We came to talk about Adrian Belkoy, who was found yesterday murdered in his house. What can you tell us about him?"

"Well, he was an incredible student. What a tragic loss. That kid was just amazing. Easy to talk to, determined, eager to succeed, and a total scholar when it came to math. He knew more than some of my teachers."

"It is a shame," replied Melinda. "We are hoping to possibly question some of the students and staff to find out any helpful details."

"Let me know what you need and you got it."

"Why don't we begin with the teachers. Hopefully, they can provide some information on his friends and things like that. Then we can go from there."

Principal Monk buzzed his secretary and asked her to retrieve a copy of Adrian's schedule. He had been taking an advanced calculus class, chemistry for college, seminar English, and an intensive senior project involving number theories and probabilities. For an All-American, Melinda and Fredric were duly impressed with his accolades and achievements. It was also mentioned that he planned on matriculating at Virginia Tech the upcoming fall on a full paid football scholarship.

"Wow," said Fredric, "that kid had it all. Who would want to kill him?"

"Would we be able to search his locker?" Melinda interjected.

"Sure. I will have to accompany you, but you have full access. We just have to wait a few minutes for the hallway to clear out."

When they arrived at locker 237, it was covered with cards and flowers, like a makeshift memorial for Adrian. Principal Monk had to move several bouquets and cards posted on the locker to open it. As this was happening, Melinda noticed that a few students had gathered behind them.

"What are you doing?" asked a long-haired redhead with an outfit short enough to fit a Barbie doll.

"We're just looking for information, ma'am," explained Fredric.

"It looks to me like you are desecrating the memorial we put up for Adrian. You better make sure you return it to the original condition or I'll sue you."

"Thank you for your concerns," replied Melinda with a pretend smile that only revealed her indifference to the student. "Sweetie, we are looking for any information to find the person who did this to your friend. I would never do anything to ruin a memorial for anyone. Second of all, you can't sue me for anything I am doing here."

"Janet, do you have a pass to be out here?" interjected Principal Monk.

She scrunched up her lips with the air of a pretentious snob. "Yes, I do, and it is right here, see?" With that, she waved a lav pass in front of his face.

"Well, let me see that, Janet. It says here that you left the room at nine-fifteen. It is now nine twenty-five. What have you been doing for the last ten minutes? I doubt it took you that long to go to the bathroom seeing you should be in English class, which is two doors away from the lav. Are you having some kind of excretory problem we should be aware of?"

Janet's cheeks turned red. She wanted to miss some class time by going to the bathroom. She never really went to the bathroom, but what student does? Most need to take a break so class goes by faster or find a chance to sneak a kiss from a loved one. Principal Monk handed the pass back to Janet. "I think you better return to class or it might be detrimental to your grades and discipline record here at Roxborough High."

Janet walked away quickly and they turned their attention back to the locker. "Hassie, do you have any gloves on you?" Melinda asked.

"No. Should I go get a kit?"

"Not necessary, I have some in my pocket. Another lesson for you: Always carry some spare gloves with you. I grab a fresh pair at the beginning of every shift. All right, let's see what we have in here."

Melinda began to sort through Adrian's locker. She cautiously placed the items on the ground as she inspected them. There was a gym uniform and a pair of mud-covered sneakers. "Hassie, we'll need the kit from the car. I'm going to bag these kicks for a soil analysis. It might be nothing, but it might place him somewhere important."

She continued her search through the locker. His notebooks, arranged by subject, were neatly organized on the shelf. The biggest of the notebooks was titled "Senior Project." As Melinda leafed through it, she noticed equations, number theories, citations from books, web sites, names of University professors and their contact information, and results from solved equations. It was similar to the beginning of a doctoral dissertation, minus the literature search and statistics. His penmanship was impeccable and his notes thorough. The next notebook Melinda pulled out was labeled "Statistics." On the first page was a beautiful drawing of poker chips, slot machines, dollar signs, and sports teams. "This is some serious bling bling," noted Melinda.

By this time, Fredric returned with a crime scene kit and a camera. He looked at the drawing as Melinda was trying to make some sense out of it. "This is some beautiful artwork. I wonder if he drew this himself. There seems to be the name of something weaved into this picture. It looks like it says Dpater Club. Hassie, what's it look like to you?"

Fredric studied it for a moment. "I think it says Drater Club. That p looks like an r. Anyway, what is a Drater Club?" he asked bewilderingly.

"I've never heard of it," replied Principal Monk. "Turn the page and find out."

Melinda perused the first few pages of the notebook to discover lists of numbers along with names of sports teams, their opponents, and game scores. After each game, there were more numbers and a W or an L. It suddenly dawned on her like a tiny light shining through a diaphanous hole in a vampire's coffin. "These are gambling entries. He's writing about how much he bet on each game, who won, and how much he came out with. He has spreads, picks, names, and numbers of people. I wonder if he was a bookie? Principal Monk, you said he was great with numbers, right? Could he have been a bookie?"

"I highly doubt it," replied Principal Monk. "I wouldn't expect something like that from him, but I didn't know him well enough to say he could be a bookie. Kids don't surprise me these days."

Just then, Melinda's cell phone buzzed. While she was talking, Fredric started leafing through the notebook, as Principal Monk looked over his shoulder for any familiar names. One name did strike a cord: Carmine DiGuiseppe. He was an intelligent student who was well on his way to life as a con man: slick, smooth, sly, and charismatic enough to talk his way out of anything. Principal Monk had suspended him several times for offenses ranging from harassment to extorting money from students. Principal Monk knew him well and had no doubt about his involvement. "Detective Hassloch, I know this kid. He goes here, and I would have no qualms about interrogating him. He is one sneaky sucker."

Melinda finished her call. She heard the last few words come out of Principal Monk's mouth. "Did you say you know a name in there?"

"Yes. Carmine DiGuiseppe. You definitely need to speak with him. Let me call the office and see where he is right now."

Principal Monk radioed the office on his walkie-talkie to ask the whereabouts of Carmine. While waiting for an answer, Melinda conferred with Fredric about her phone call. "That was Dr. DiGregorio on the phone. He's going to begin the autopsy of Adrian's body in an hour. We need to be there for that one. I know it's your second one in two days, but crime never takes a holiday. Good thing you have a strong stomach, Hassie. I trained this guy once who had some kind of stomach problem. He couldn't handle any autopsy without throwing up. The weird thing was, he didn't mind the gruesome crime scenes, and once he threw up, he was all right. That guy left the department after one year and moved to Cutbank, Montana, to become the head sheriff of the town. I haven't heard from him since. My brothers used to dissect everything they found, so that stuff doesn't bother me."

Principal Monk finished his conversation with the office. "He isn't here today; shocker. He misses a lot of school. Let's head back to the office and I'll give you his address. Did you want to talk to any of the teachers or students today?"

"I'd love to," replied Melinda, "but we have the autopsy for Adrian in an hour. Let's clean everything up while Hassie gets some pictures and bags this stuff. This Carmine DiGuiseppe kid, do you think he could have been involved?"

"You never know with that kid," said Principal Monk. "He usually has malicious intentions, but not to the point of murder. He always gets right to the edge where he comes up with some excuse to get himself out of trouble. The times that I've suspended him, it was because of accusations from other students, and Carmine never really said he *didn't* do something. I'll pull his file for you. Take it with you and copy what you need."

The threesome wrapped up their visit, and Melinda and Fredric left with the evidence and file on their new possible suspect, Carmine DiGuiseppe.

✋ ✋

Vocabulary

Control center   A central location that is established to permit or deny persons into a crime scene.

Entry point   The place or area where a criminal may have entered. It also refers to the place or area where designated people enter the crime scene.

Exit point   The place or area where a criminal may have left the scene. It also refers to the place or area where designated people exit the crime scene.

First responder   The responding officer who is the first one to arrive at the scene of a crime.

Primary incident site   The place or area where the actual crime is supposed to have occurred.

Protocol   The rules or conventions of correct behavior to be followed on official occasions.

Sensory observations   Observations that are directly related to the five senses (sight, taste, touch, sound, smell).

Victim   The person whom against a crime or offense has been committed.

Witness   A person who is present during a crime.

Background Information

The first responding officer is the first person to arrive at the scene of a crime. First and foremost, he or she must attend to any victims in need of immediate attention. The first responder might also secure witnesses and keep them from leaving the scene of the crime, and, of course, the crime might still be happening. A first responder might have one of these particular situations or all of them at once. Whatever the case, instantaneous decisions must be made.

Imagine being a first responding officer and arriving at the scene of a crime, which is a shooting on a busy street corner. One person lies on the ground with life-threatening wounds. Five feet away is a fleeing suspect, and shells from the firearm are spread on the ground. How can people be kept from walking into the crime scene? Do you radio for help and chase the bad guy? Should you administer CPR? This is just a small example of what real police officers face while on the job.

There are some basic responsibilities of the responding officer. When he or she arrives, no definitive assumption can be made whether or not the crime is still occurring, so the officer should assume the scene is still active and dangerous. The first item of importance is to call for assistance, an ambulance, firefighters, or other personnel. Next is identifying any persons, whether they are living, dead, injured, or intoxicated from drugs, alcohol, or both. If cognizant people are available, try to find the one who called police. When there is a crime with much violence, it may be difficult for the officer to maintain control because the crime becomes susceptible to attention by higher officials as well as emotionally charged victims, witnesses, and curious neighbors or bystanders. They must be kept at bay to prevent compromising of evidence.

Following group organization, entry and exit points should be established and secured to keep anyone from entering the area and contaminating any possible evidence. Another noteworthy item is recognizing and recording any sensory observations that pertain to sight, smell, sound, taste, and touch. Sometimes these sensory details are apparent upon the arrival of the responding officer but disappear when backup arrives.

Sometimes emergency personnel and detectives are delayed in their response to the scene. If so, the first responder may have to administer medical care to victims or other people involved. When the medical team does arrive, lead them along a designated path that will cause the least amount of disruption to the crime scene. It is imperative that the medical team be made aware of the need to preserve evidence. It may consist of bullets or the victim's clothing, or other pieces of crucial evidence. Most medical staff is trained to handle these situations. It should also be noted what hospital victims are transported to; someone will eventually have to visit them for questioning about the crime. If victims or witnesses at the scene do not need medical attention, they should be interviewed on the spot for information. If there are several witnesses, it is desirable to separate them so they do not corroborate stories that might prove false.

After the professionals arrive and anyone in need of medical assistance has received attention, it is time to secure the scene and set up a protocol. Up to this point, the only people within the actual crime scene are the medical personnel and any detective or investigator who needs to be there. Most times, police officers who are not trained in the area of forensic investigation are not permitted to enter the scene. After witnesses have been placed in secure areas, the scene can be properly taped off or barricaded with the appropriate material. Working outward from the primary incident site, secure all sites of possible entry and exit along with any vehicles that may be associated with the crime. A control center should be set up, which is a central location for permitting or denying access to the crime scene. One officer should be in charge of this job. Everyone who enters the crime scene must sign a log with their name, position, date, time, badge number, and reason for being there. Also, whoever enters the crime scene should have their prints in the AFIS system. To err is human, and sometimes prints end up in places they are not supposed to.

Another concern is the media. In a high-profile case, the media is sure to arrive quickly with their cameras rolling. To keep them from becoming intrusive, barricades should be set up around the outside perimeter of the scene and guarded by other officers.

As detectives or technicians begin to examine the crime scene, the first thing to be identified is any possible DNA evidence. This may be anything from blood to wine glasses to cigarette butts. Before this evidence is touched, where it was found should be documented. A sketch can be drawn, a picture taken, notes written, or a video recorded. It is recommended to take all of these measures to prevent any questions in the courtroom. One item that is paramount when taking photographs is to place a ruler next to the piece of evidence to give an approximation of the size and length of the evidence. After documentation, the items can be packaged and taken to the proper area for temporary storage.

Photography is one of the most essential tools for the crime scene investigator. Integrity of evidence can determine whether a person accused of a crime goes free or goes to jail. All evidence that is photographed must be done in an unaltered condition. Exceptions would be any piece of evidence that is involved with an injured person that must be removed. If objects are removed or the position changed, the item may not be admissible in court and the case will be lost. If the item must be removed for life or death purposes, it should be well documented in the report.

Every crime scene should be photographed as completely as possible. Many pictures should be taken and from different angles. A wide-angle lens might be used to shoot an entire room, where a one-to-one lens is used to show the actual size. A macro lens may also be used to shoot close-up pictures of wounds or other evidence. Pictures should be taken of the general area of the crime and any adjacent areas that may be important to the crime. Points of entry and exit should be photographed. If the scene is indoors, the entire room should be photographed starting with the point of entry. Surrounding rooms should also be photographed for reference. If a body is somewhere within the scene, it must be photographed from far away to establish a reference point of where the body was found. Then closeup shots can be taken. After removal of the body, the underlying surface should be photographed to expose other potential evidence.

Forensic Implications and Jurisprudence

Collecting evidence must be done with great care, for contamination or any misinformation can compromise it. What kind of evidence is found determines what kind of packaging is used. Some evidence bags are made of plastic with a piece of cellophane paper that can be removed to expose an adhesive strip, which is virtually impossible to open without showing signs of tampering. If the adhesive strip is not available, evidence collection tape should be used. It is composed of a flimsy, sticky tape that can be signed by the investigator. It rips with ease, so it is extremely difficult if not impossible to pull it off without ripping or distorting the tape.

Eventually, it will be opened at the lab from another place to keep the tape from ripping. When the next person is finished looking at the evidence, it is sealed with the same kind of tape and signed.

Any piece of evidence that may contain DNA is packaged in paper. This prevents any bacteria or condensation that could grow and degrade the sample. For firearms like rifles or other long-barrel guns, special boxes and bags are made that can be purchased through any crime scene supply company. No matter what evidence is found and collected, all of it needs specific information: case number (if applicable), item number, officer who collected the evidence, the police department, their contact information, victim's name (if applicable), location of the evidence, and any serial number or any outstanding markings that are specific to the piece of evidence. If possible, a detective or technician may inscribe his or her initials in the actual piece of evidence. This holds tremendous power in the courtroom when a defense lawyer asks how the detective knows that it is the same piece of evidence found at the scene.

When collecting evidence such as soil, blood, glass, hair, and fibers, a control sample should also be collected for comparison in the lab. Although most investigators can identify and collect pertinent crime scene evidence, some forget the importance of having a control sample to submit to the lab.

When the evidence is taken to the lab, a protocol is followed as well. The officer who delivers the evidence must sign in with his or her contact information, the items being dropped off, and their condition. In many cases, evidence is sent through the mail because crime labs are not ubiquitous in every county. Some restrictions prohibit the use of mail for submission of biological fluids and other hazardous materials, so they must be delivered by hand.

Once the evidence is handed over, it is beneficial for the officer to speak to the technicians examining the evidence. The lab should be provided with a brief description of the case history. This helps scientists to analyze specimens in a logical sequence and make the correct comparisons. After the scientist is finished examining and testing the evidence, it should be placed in the same package in which it arrived. It should be resealed with evidence tape and signed by the scientist or technician completing the analysis.

Along with filling out evidence bags, some crimes require specific reports and paperwork to be completed. These official forms are eventually submitted as evidence in court. When an investigator fills out a report at the scene, he or she need not worry about grammar and mechanics. However, when the report is to be prepared for court, spelling and grammar are of the utmost importance. Every "i" should be dotted and every "t" crossed. Defense lawyers love nothing more than an incompetent investigator who does not have a solid command of the English language.

Now that the initial legwork is completed, investigators can begin to organize their information and actually investigate the case.

Hassie's Second Date with Death

The room was chilly and the fluorescent lights buzzed overhead. The stainless steel tables were scrubbed clean and the faucet had a slow drip that pierced the silence every five seconds. The morgue was a desolate place, but only in the depths of the hospital basement did the dead speak to the pathologist to unlock the mysteries of death. Melinda and Fredric stood silently glancing around the room, waiting for Dr. DiGregorio and Catherine, his deenir (which is a fancy word for assistant). The doors squeaked as the two entered the room wearing blue hospital scrubs. Dr. DiGregorio was in his usual chipper mood as he extended a hand. "Good afternoon, Detective Monsternick. Hassie, how are you holding up after yesterday? Stomach staying with you?"

"Yeah," replied Fredric, "I have some strong nerves. I would imagine this one won't be as putrid and stinky as yesterday. Nothing like a rancid body for my first autopsy experience."

"Everything will be easier from here on out," offered Dr. DiGregorio. "Catherine, let's go grab the body and get some gear on. You guys can put those gowns on over there and some head wear. I wouldn't want you getting a brain chip in your eye when Cat fires up the Stryker saw."

Melinda chuckled to herself. Dr. DiGregorio had a twisted sense of humor that most normal people would not understand. Death seemed to have an unusual way of affecting people. For most in the business, a macabre mentality of deranged comedy helped to look at the body as a piece of science. Desensitization was essential when dealing with human life, or lack thereof.

Cat wheeled out the gurney to the autopsy table. Dr. DiGregorio snapped on his gloves and pulled out a mini tape recorder, as was customary for a cautious pathologist concerned with his work. "Hassie, since this guy isn't as decomposed as our body yesterday, we can run an extensive external examination of the body, which might provide some clues as to how this kid died. Gosh, I hate saying the word 'kid.' These are the hardest ones to do. I try not to think of my own children, but sometimes it's inevitable. So, let's look at the various areas of trauma and try to discern which wounds happened first."

Dr. DiGregorio began to examine the body with a magnifying glass. He poked at a few of the bruises and the skin blanched momentarily, then returned to its now deep purplish color. Next he examined the slash on the neck and turned his tape recorder on. "The patient is an 18-year-old male Caucasian named Adrian Belkoy," he bellowed as he read the police report. "There is trauma to the chest that appears to have occurred from a blow from a blunt object. There are no apparent wounds to the lower extremities or arms. The cut on the neck is through the jugular vein. Upon closer magnification, it appears that the skin is serrated." Dr. DiGregorio looked up momentarily. "Did you find a murder weapon?" he asked.

"We think so," said Fredric. "I found a serrated knife covered with blood in the kitchen."

"That's interesting. Cat, get the camera and take some close-up pictures of the wound. Don't forget the rulers." Cat pulled out the camera and attached the macro lens for some close-up shots.

"There is a clear cut through the vein. Were there a lot of arterials around?" questioned Dr. DiGregorio.

"Yes," answered Melinda, "there were several in the kitchen along with some passive drops next to the sink. I think the perp might have been standing there for a few seconds. There were two puddles of secondary spatter; you know, a bunch of drops in one spot. The perp couldn't have been standing there for more than five seconds."

"Wow," interjected Cat. "I would imagine the place was a mess."

"That's putting it lightly," responded Melinda. "There was blood everywhere. We had a high-impact spatter against the wall at the other end of the kitchen. I would venture a guess that it was from the gunshot. The wall looked like red sandpaper. It wasn't as much as one would expect, though, probably from the lack of blood from the cut throat."

Cat grimaced. She had been working with Dr DiGregorio for 15 years now and still did not like the thought of any blood spilling from young, promising children. "Doc, are we ready to check out the gunshot wound?"

"Yes, let's take a look." He turned on his tape recorder again as he examined the gunshot wound to the head. "It looks as if this is a direct-contact gunshot wound. There is starring and no stippling on the scalp."

"What does that mean?" inquired Fredric.

"Well, Hassie," Dr. DiGregorio stopped his recorder as he spoke. "It means that the barrel of the gun was in direct contact with the skin on the head. When a gun—more particularly in this case, probably a pistol or handgun—is fired up against the skin, the gases from the gunshot actually go inside the head and rip the skin from the bone. It is so forceful that it rips and tears the skin, leaving a starlike shape on the head. If you come over here I'll show you."

Fredric looked on as Dr. DiGregorio explained that no soot could be seen outside the gunshot wound, for when the barrel is not in direct contact with the skin, gunpowder shows up as tiny, black, stipplelike marks around the wound. Then Dr. DiGregorio took out a swab and sampled from around the gunshot wound. He labeled it and placed it in a vial to be shipped to the crime lab. "If you found the gun immediately after the gunshot, you would be able to see fragments of bone and tissue inside the barrel of the gun," he said. "This is called the blowback effect. As the gases escape, a vacuum is created in the barrel and sucks back in, hence the fragments of skin and bone. A stereomicroscope works great for that."

Dr. DiGregorio continued, "I don't see an exit wound, so the bullet must still be inside his head. Cat, let's get an X-ray. Another thing, Hassie, when a bullet goes through anything, sometimes it stays together and sometimes it shatters. We'll be able to tell for sure after the X-ray. If the bullet has broken into fragments, it might look like snowflakes in the head. We call this metallic snow. If it's intact, we'll be able to retrieve it and reserve it for testing and the courtroom. I'm sure Monsternick will tell you all about firearm identification and ballistics."

Cat wheeled over the body for several X-rays of the head. As they waited, Dr. DiGregorio offered his thoughts about the two bodies that were related by a single gunshot wound to the head. "Monsternick, you could have a serial killer on your hands."

"I know," she said, "I have been thinking about it all morning. But this guy had his throat slit first. I would imagine he was half-dead when he was shot."

"We'll be able to tell once we look at the amount of hemorrhaging in the brain. Same execution-style gunshot wound, though, two male bodies, one a high school student, the other unidentified."

"Not for long," Fredric interjected. "We should have a DNA profile in a few. I'm hoping we can put an end to the missing high school kid over in Roxborough."

"Oh, you mean Mindrago?" replied Dr. DiGregorio. "Yeah, what a shame. That kid was the next All-American LeBron James. You know, you have a good shot with that body we had yesterday, no pun intended. From the looks of his skull, the fissures weren't fully closed, so the victim could be anywhere from eighteen to twenty-eight years old. Not to worry, though, your DNA test will tell all."

"We have Arthur Ying doing the work for us at National Medical Labs," said Melinda.

"Those guys are fantastic. We send a lot of work their way."

"Okay," said Cat as she locked the wheels on the gurney. "Should we open him up?"

Dr. DiGregorio pulled out a fresh scalpel. He made the typical Y incision across the chest that ended in the pubic region. The skin separated like silk as the scalpel cut through layers of muscle, tissue, and adipose fat. As Cat inserted the clamps to open up the chest, she and Dr. DiGregorio immediately noticed the *situs inversus:* a rare medical phenomenon in which the organs are reversed from the normal human anatomy. The pulmonary system is reversed so that the left lung is larger than the right lung. Also, the liver and gallbladder are located on the left, while the spleen and stomach are located on the right. All of the other organs in the gut are opposite as well.

"Wow," exclaimed Dr. DiGregorio. "We have a case of *situs inversus* here. Everything is opposite of what it should be in a normal human. You can see the difference in the organ placement. This is the first one I have ever seen in my twenty-five years as a pathologist. You are witnessing a true anatomical phenomenon. It doesn't have any comprehensive effects on how the body functions, but it is an oddity."

Melinda and Fredric looked at each other, impressed with their new knowledge, hoping to disseminate their information back at the office with their colleagues. Melinda now focused her attention on the bruises. "Doc, what about the bruises on the chest?"

"Well, let's peel back the epidermis and see."

As he exposed the targeted area, the surrounding tissue was a dark reddish-brown color; a sure indication of hemorrhaging from some kind of instrument. "We have some serious trauma down to the subcutaneous level. What happens is that the blood vessels break and they leak. It stains the muscle and tissue. These are antemortem because there is hemorrhaging. It was probably caused from punches or severe blows from a blunt object."

Dr. DiGregorio was sure to explain his terms so Fredric would understand the medical terminology. Being the new guy, he couldn't retain every piece of information he had to read, so explanations and brief encapsulations helped to expand his background.

"All of the organs seem to be intact with no visible trauma," said Dr. DiGregorio as he investigated the visceral area. "See, Hassie, this is the visceral sac. Basically, it's a sac that holds all of your organs together. Visceral just means your abdominal area. We have to use the fancy terms for the paperwork to keep it consistent for jurisprudence in the courtroom."

Dr. DiGregorio slowly began to take out one organ at a time. He started with the stomach. The first thing he did was have Cat weigh it and record it. Then the stomach was turned inside out and emptied into a container labeled "gastric contents." This would tell what the victim's last meal had been, and possibly how long it had been in the stomach. It might not be significant in this case, but in cases of poisoning and determining time of death, gastric contents were crucial. Upon reaching the heart, Dr. DiGregorio found it normal and healthy. There were no other discrepancies in any of the organs, but they were still examined.

"Okay, guys," began Dr. DiGregorio, "there are no problems with his organs. Aside from the *situs inversus*, he was quite a healthy kid, which doesn't surprise me, seeing he was a stellar athlete. Let's take a look at the throat."

Dr. DiGregorio dissected the neck area, investigating the spinal area, pharynx, larynx, and other esophageal areas. "This slice through the jugular was made with one fluent motion. The hyoid bone is intact, so we can rule out manual strangulation. There is no trauma to the other areas. He probably had about six or seven seconds before the blood drained out from the brain. What a horrible way to go, although not as painful as other egregious acts committed against the body. Cat, get the Stryker saw ready so we can look at the brain."

At this point in the autopsy, Adrian Belkoy was not in their vocabulary. Dr. DiGregorio, Cat, and Melinda were so far immersed in the autopsy they were more concerned about finding evidence rather than pondering thoughts of a young talented athlete with a bright future. Their job was to be a medical expert, scientist, and investigator to ensure all possible evidence was taken into account to bring justice to the now-deceased Adrian. Fredric, on the other hand, was not yet adept at distancing himself from victims of crime at this point in his development as a budding detective. He had constant thoughts about the crime, the loss of life, and the effects on those close to Adrian.

Cat went to work with the saw and made a circular cut from ear to ear and across the top of the forehead. The skin peeled off like a mask, exposing the gunshot wound in greater detail. Dr. DiGregorio once again turned on his tape recorder. "Upon removal of skin from the head, there is obvious gunshot residue visible around the entrance wound. The entrance wound is identified by the concave indentation in the occipital bone. Hemorrhaging is minimal due to the loss of blood from the cutting of the jugular vein."

He clicked off the tape recorder again. "Cat, we need to take out the brain to trace the path of the bullet, but first, let's see the X-rays." He placed them on the light board, where they showed an intact bullet awkwardly placed in the lower portion of the skull. "Looks like you'll have another piece of evidence, Monsternick," he said confidently. "Okay, Cat, take it out."

After recording its weight, the brain was placed on the table. Dr. DiGregorio pulled out a serrated knife that looked similar to a bread knife. He sliced the brain in half-inch slivers, exposing the path of the bullet. "It looks like it made it down to the hippocampus area." Using his gloved fingers, he carefully retrieved the bullet, placed it on a piece of cloth, and passed it over to Melinda.

"This looks like a .45 caliber, she said. "It's in remarkable condition. If we can find the gun, we should be able to get a match. Hassie, remember I was telling you about lands and grooves on a bullet?"

Fredric's mouth opened like he was going to say something.

"Don't answer," said Melinda. "It's a rhetorical question, meaning you don't have to answer it. We'll have a nice profile from this one. The comparison microscope and digital imaging will confirm the identification. Doc, nice job."

"Thanks, Monsternick, I must say that you have a nice sample for the lab. I hope you catch this psycho before I see another body."

"We'll do our best, Doc," smiled Melinda. "Justice is always waiting to happen. We just need to be at the right place at the right time. Hassie, what are your thoughts?"

Fredric stroked his chin pensively. "I think there is a strong probability of a connection between the two bodies, but we have no suspects or motive. I guess once we get the results from yesterday's John Doe, we'll be one step closer to finding a perp."

"Right," said Melinda. "Doc, we have to get back to the office to fill out some paperwork; the true motivation for my love of police work."

"Yeah, right," chuckled Dr. DiGregorio. "Paperwork is just another way to keep us from catching a bad guy or examining another body. I'll have your reports in a day or two. Until then, don't send me any more bodies. Cat here needs some time to type up my reports, and I have a three o'clock tee time with the DA this afternoon. Take care and talk to you soon. Hassie, I hope you're becoming more attuned to gunshot wounds and *situs inversus*."

"I sure am," replied Fredric. He shook hands with Dr. DiGregorio and thanked him for his expertise and tutelage.

As they made it to the garage, Melinda spoke first. "Hassie, I can see you're still learning the ropes of building up your defense mechanisms for dealing with this stuff, so make sure you talk to me or schedule some time with the department shrink. She made my life easier when I was starting out."

"I know," he said. "I can't stop thinking about this guy for some reason. I put a name to this kid and he's incessantly pervading my thoughts. I'm going to do everything in my power to catch this son-of-a-gun and put him behind bars. You know, I don't have kids, but I love my niece and nephews. If anything ever happened to them, I don't know what I would do." Fredric's eyes welled up with tears, as Melinda grabbed him and gave him a hug. Right there in the parking garage, Fredric let his emotions out, soaking Melinda's shoulder with his tears. She had done it before as well as other detectives around the world. It was all part of growing into the role of an investigator. Fredric was surely on his way.

Vocabulary

Algor mortis   The temperature or cooling process of a body after death.

Antemortem   Any event occurring before death.

Autopsy   The examination of a deceased body to determine the cause, manner, and sometimes mechanism of death.

Cause of death   The immediate reason for death in a human.

Contusion ring   A portion of stretched skin around the entrance wound caused by the rotation of a projectile.

Coroner   An elected official who is permitted to conduct autopsies. Does not require a medical degree, but has medical training.

Demonstrative evidence   Any piece of evidence viewed by the jury to strengthen or weaken an argument or claim.

Entrance wound   The area of the body where a projectile enters.

Exit wound   The area of the body where a projectile exits.

Forensic pathologist   A doctor who has received years of training and is certified to conduct autopsies and testify in court.

Gross anatomy   The study of body parts that are visible to the naked eye.

Homicide   The intentional killing of one person by another.

Livor mortis   The outward appearance or color of a deceased body.

Manner of death   The reason of death, which can be homicidal, suicidal, accidental, natural, or undetermined.

Mechanism of death   The object, whether it be known or foreign, that was used to cause a death.

Medical examiner   A physician officially authorized by a government unit to ascertain the cause of death. Unlike a coroner, the medical examiner is always a physician.

Pathology   The scientific study of the nature of disease and its causes, processes, development, and consequences.

Postmortem   Any event occurring after death.

Rigor mortis   After death, the stiffening of joints, muscles, and tissues in the body. It can last anywhere from 12 to 36 hours.

Serology   The branch of medical science that deals with serums; especially with blood serums and disease.

Smudging   A blackened area that results from gunpowder smoke and gases.

Starring effect   A star-shaped wound that is produced from exploding gases under the skin.

Stippling   Tiny dots around the surface of an entrance wound from burned and partially burned bits of gunpowder.

Toxicology   The study of the nature, effects, and detection of poisons and the treatment of poisoning.

Velocity   A vector measurement of the rate and direction of motion.

Visceral area   Having to do with the viscera, which are the soft internal organs of the body, including the lungs, heart, and organs of the digestive, excretory, reproductive, and circulatory systems.

Background Information

The autopsy is one of the most crucial areas of forensic science. It pertains to determining the cause, manner, and sometimes mechanism of death, which is then stated on the death certificate. Not all deaths require an autopsy, which is expensive and time consuming. The following is a list for autopsy requirements.

Automatic Autopsy

- Homicide and suspected homicides
- Deaths that are suspicious
- Unknown cause or manner of death in cases where the death is sudden, especially in children
- Hit-and-run victims
- Inmates who die in prison
- Deaths of individuals who operate public transportation
- Deaths that involve potential hazards to the public

Possible Autopsy

An autopsy might be required in the following cases:

- Death that is likely to trigger legal proceedings, such as a wrongful death
- Unidentified bodies
- Motor vehicle crash victims where criminal charges are likely

- Prominent or notorious deaths
- Suicides
- Natural but unexpected deaths

Items That May Rule Out Autopsy

An autopsy might not be performed in the following cases:

- Medical history accounts for the death
- Age of decedent
- Family objections
- Cost
- Religion
- No pathologist available

In the past, pathology was the study of structural changes caused by disease, but today, the field has expanded to include the studies of disease insofar as it may be investigated by laboratory methods.

A forensic pathologist is expected to provide the following information during an investigation:

- Determine a diagnosis of apparent cause of death
- Determine whether the death was natural, accidental, homicidal, suicidal, or undetermined
- Provide an estimation of the time of death
- If there are any wounds on the body, determine the instrument that was used to make them
- Make clear distinctions from deaths that are homicidal, suicidal, accidental, or natural
- Establish the identity of the deceased if not already done

The medical examiner system was established in Massachusetts in 1887. Some states have central offices while others have local examiners who work under the state medical examiner. The medical examiner is often vested with some aspect of judicial function in that he or she may conduct an inquest as to the cause of death, manner, and mechanism. The primary role of the examiner is to perform the actual autopsy and conduct microscopic tissue examinations. He or she must also order the appropriate chemical, toxicological, serological, or bacteriological analyses. All of this information must be interrelated with the data from the investigation.

There are two general fields of pathology: anatomic pathology and forensic pathology. If one desires to combine the two, he or she is expected to spend a minimum of two years of training in each field prior to the examination for certification.

In some states, the coroner system is used. The coroner is an elected official, and in some instances, need not be a medical doctor. More recently, the coroner does possess medical training. This system stemmed from the old one in England, where the "crowner" was a man who worked for the king and who was in charge of collecting death taxes from the families of those who had passed on. Suicides were punishable by law, in which case families were expected to pay a fine. As times progressed, the position of coroner was held by several different people. In some instances, woodworkers became coroners because they could make the coffins for people. They could be purchased with other items in the store like a table and chairs if one desired.

Technically speaking, the word *autopsy* is defined as "a personal observation." The etymology of the word comes from the Greek *auto*, meaning *self*, and *opsy*, meaning *eye*. According to the Oxford Dictionary, the autopsy first meant, "seeing with one's own eyes, personal observation or inspection." Its purpose is to observe and record the minute and gross anatomical peculiarities discovered on or within the dead body. The results of the autopsy provide the real evidence used to predicate the medico-legal opinions about the cause, manner, and possible mechanism of death.

Throughout the autopsy, copious notes must be taken for future reports and depositions that are to be used in court. A sketch should be provided of the pertinent features of any demonstrative evidence crucial to the investigation. Photographs are commonly taken at autopsy and become a significant factor in the courtroom. However, before pictures make it to the courtroom, they must be deemed admissible by the judge.

The typical autopsy report includes a general description of physical features of the deceased (gender, age, race, and so on). The signs of death are also noted and help establish the approximate time of death. These include rigor mortis, livor mortis, and algor mortis. Rigor mortis is the stiffening of the body due to a breakdown of proteins. They form chemical bridges, which causes the stiffness in the body. It usually lasts anywhere from 12 to 48 hours. Livor mortis is the color of the body. As it decomposes, color changes are seen and can sometimes be affected or changed by use of drugs or other chemicals. Algor mortis is the temperature of the body, which is more pertinent when the body is found after the first several hours of death.

Forensic Implications and Jurisprudence

The autopsy begins with an external examination of the body. The pathologist examines the head, trunk, extremities, and genitalia for any discrepancies or irregularities. An examination of external wounds must also be charted, including the pathway of those wounds through the body.

Sometimes it is necessary to include X-rays of the deceased. X-rays are taken in all firearm injuries because the bullet may be retrieved from the body as evidence. If the bullet has exited the body, fragments may be available for forensic examination. X-rays are also useful when examining stab wounds, because part of the instrument may have broken off and be embedded in the muscle or tissue. Decomposed bodies and fire deaths also require X-rays to make identifications through bone or dental record comparisons.

Next, a Y incision is made starting at the two shoulders and ending in the pubic region. The sternum and ribs are cut and opened to expose the visceral area and upper portion of the chest. Examining the inside of the body is usually the most time consuming part of the autopsy. In the past, only the brain, stomach, heart, and possibly the kidneys were examined. Today, the internal examination looks at the organ systems, heart, major vessel systems, liver, stomach and its contents, small intestines, rectum, genitalia, neck, brain, and the spinal cord. Some organ specimens are taken after being weighed and examined for any irregularities. They are then sent to the proper laboratories for analysis by a trained professional. As the pathologist completes the examination, he or she should also be looking for signs of mechanical or dynamic injury. A search should be made for conditions such as heart disease, lung disease, or disease of any of the major body systems. Disease may be caused by the presence of hereditary, congenital, infectious, cancerous, metabolic, or toxic abnormalities rather than trauma. It is of the utmost importance that the pathologist be able to distinguish between structural changes produced by trauma and those produced by disease.

When reporting the cause of death, the pathologist provides the immediate cause and any additional causes of death. For example, if the victim has been stabbed in the stomach and dies from blood loss, the blood loss, also known as exsanguination, is the immediate cause of death. The puncturing of the stomach would contribute to the cause of death and the knife itself would be the mechanism.

The manner of death is important because it determines whether charges can be brought into play, and a court date will be most likely. It is homicide if someone other than the decedent was the source of the cause of death. When the manner of the death is accidental, a person causes a death without intending any harm. This is difficult to prove in a courtroom and many cases have been argued over this type of death. A suicidal manner of death is when a person intentionally causes his or her death. The last manner of death is natural, which means the person died from natural causes in the environment. One exception here is the undetermined manner of death. This arises when the pathologist cannot determine the actual cause of death. The mechanism may be known, but it does not dictate the immediate cause of death, and therefore it is undetermined.

Gunshot Wounds

When dealing with gunshot wounds, the pathologist examines the wound and determines the entrance and exit points, the nature of the weapon and projectile used, the number of gunshot wounds, their location, the possible distance, direction of fire, and in the case of foul play, the position of the victim and the shooter. The latter usually proves to be the most problematic.

Extreme caution must be used when retrieving any foreign object from the body. Forceps and probes can cause striations and damage that can compromise the evidence. The pathologist determines the extent and course of the wound, whether it was fatal, whether it was self-inflicted, and whether it occurred antemortem (before death) or postmortem (after death).

In terms of ballistics, the greater the energy of the projectile at the moment of impact, the greater the destruction to the tissue. According to physics, the striking energy of a projectile is the product of its mass or weight multiplied by the square of its velocity. Velocity is the most important factor because it affects the impact and damage to the body. Handguns are not usually as damaging as a high-powered rifle because handguns have a lower velocity speed. A shotgun may cause severe damage because the scatter of shot entering the body can make multiple entries and wreak havoc within the tissue and muscle.

The type of weapon used determines the type of wound created. Also, the area of the entrance wound varies because of body composition at various places. The distance is another constituent that determines the size, shape, and damage of the projectile. As a bullet contacts the skin, it is moving at a high speed and rotating because of the rifling in the barrel. At the point of entry, the bullet perforates the skin and pushes it, forcing the skin to stretch. After entry, the skin retracts, and for most gunshots that are not direct contact, will lose the original shape of the projectile. The wound may roughly correspond to the caliber of the gun, but it would not be considered a true measurement.

A direct-contact gunshot wound results when the muzzle of the firearm is placed in a position where it is touching the skin of the victim. A close-range gunshot wound is one in which the muzzle is held approximately 1 to 20 inches from the point of entry.

Close-range gunshots are defined with several apparent patterns. One that may be observed is the scorching of hairs and skin, which results from the flame discharge and hot powder gases that come in contact with the body. The extent of scorching and burning depends on the type of firearm used, approximate distance, type of powder, and the surface of the entry point. Sometimes a blackening, also called smudging, occurs from the powder, smoke, and gases. It is discovered internally if it is a direct-contact shot. Another indication of a close-range gunshot is a contusion ring, sometimes called an abrasion collar, which is a portion of stretched skin around the entrance wound. It is caused by the rotation of the bullet against the skin prior to penetration. In a direct gunshot, this is not easily identified.

Stippling is another indication of a close-range gunshot. The epidermal layer of skin is embedded with grains of burned or partially burned powder that is visible to the naked eye. It looks like a bunch of tiny dots placed around the point of entry. Direct-contact shots usually show evidence of the starring effect. When the muzzle of the firearm is held to the skin, there is no place for the gases and powder to escape. Therefore, the explosion occurs underneath the skin and forces the surrounding area to tear and rip. Most of the time the resulting effect is a star-shaped wound. If the gun is available, bits of skin and bone may be present in the barrel. These can easily be seen with a stereomicroscope. A vacuum is created when the gun is fired and actually sucks in the fragments from the body. Lastly, a muzzle imprint may be evident on the skin. This occurs in most direct-contact shots because the force of the explosion pushes the muzzle against the skin and causes hemorrhaging or bruising.

Of course, if a projectile goes through another object before entering the body, these indicators will not be seen. In this case, a search should be completed to find the object that was penetrated by the shot before it entered the body. This object will contain gunshot residue as well as a hole where the projectile passed through, which can later be matched to the firearm in question.

Even though pathology is viewed as a specialty of medicine, expert testimony is absolutely appropriate since the jury cannot be expected to possess the knowledge to interpret forensic findings. Generally, the court systems do not impose a requirement of specialization of board certification in pathology as a prerequisite to appear as an expert on the stand. This is probably because specialists other than pathologists qualify as experts with regard to studies made on human body tissues or fluids as a result of an autopsy. Many courts have upheld rulings that a qualified physician can express an opinion on the cause of death.

The pathologist is permitted to use demonstrative evidence to provide a better understanding for jury members. At the discretion of the judge, the pathologist may use photographs, color slides, infrared photographs, X-rays, charts, drawings, skeletons, and videotapes of the autopsy.

Admissibility of gruesome photographs is often the topic of debate. These photographs may contain demonstrative evidence that has a direct bearing on the investigative process and is indeed necessary to see. Photos of victims found at the crime scene are admitted because they establish location, condition, and position of the body. Other times, it may be difficult to show such photographs when family members of the deceased are present. In either case, the judge has the right to admit or deny this type of evidence.

In order to qualify as a forensic pathologist, one must graduate from an accredited medical school and complete a minimum of four additional years of specialized training. He or she must then pass an exam provided by the American Board of Pathology.

5

Carmine's College of Carnal Knowledge

"**C**ome on, I don't think you have the marbles to do it," encouraged Carmine.

"You don't think I'll do it? I'll do it, then I'll do it again just to prove I ain't scared of no one," Kevin yelled back.

Kevin Sangiovesse picked up the cinderblock and smashed it through the rear window of a brand new white 7 Series BMW. The siren immediately echoed throughout the streets of Manayunk in the still night air, but most of the city dwellers had gone to bed. Carmine and Kevin had spent the night playing cards with some classmates and a card shark named Jo Jo. Carmine brought his unsuspecting and supposed friends to Jo Jo's house so he could clean them out. Of course, Jo Jo provided the alcohol, lowering their brainwaves and making them think they had winning hands, only until Jo Jo laid out full houses aces high and four of a kinds with kings.

Now Carmine and Kevin were walking the streets of Manayunk in the early hours of Saturday morning. Carmine was somewhat sober, while Kevin was drunk and belligerent about losing his recently cashed paycheck to Jo Jo. Kevin was short the 700 dollars he owed to Carmine. Being quick on his feet, Carmine decided to take advantage of his friend's drunken stupor and entice him to throw the cinderblocks through the car window. For each window he broke, Carmine would knock $100 off his debt. In his state of inebriation, Kevin did not seem to mind as Carmine filmed him using the camera on his cell phone. He smiled for the camera as he stood in front of the car and raised the second cinderblock high above his head.

"I should get two hundred for this one, Gizep," Kevin huffed as he launched the block through the front windshield. A million shards of glass scattered all over the hood and sprinkled onto the ground. He looked at Carmine drunkenly. "How's that, you want more?"

"How about we say you owe me four hundred and we'll call it even," offered Carmine.

"You got it, Gizep."

As the police car turned the corner and hit the lights, the two of them ran in different directions. Luckily for them, the back alleys of Manayunk were barely large enough for a golf cart let alone a squad car. Their escapade ended with police cars scanning the streets vigilantly while Kevin made it to the banks of the Schukyll River to find a hiding spot in the bushes where he crashed for a few hours. Carmine was smart enough to make it home and climb through his first-floor window undetected.

When Monday rolled around, Carmine decided to bless his class-mates at school and show up for class. His white pressed polo shirt and khakis were his way of exploiting the fact he had money and could spend it as well. The profligate Carmine never wearied of indulging excessively to make himself look fashionable. Most students knew his business and stayed away from him. Those who were brave enough, or in this case stu-pid enough, to associate with him usually found themselves in some sort of trouble. There was a story floating around school that he had black-mailed a lawyer for $20,000. Carmine had seen this guy around town on several occasions with different women and wearing a wedding band to boot. Opportunity knocked, and Carmine had an easy target. Knowing what some of the married men do in Philly, Carmine made a friend in school and offered her $1,000 to get the lawyer in a somewhat precarious situation. Carmine used his Nikon camera with the telephoto lens to capture some beautiful shots of this guy making the moves on her. After Carmine told the man he could ruin his marriage and his wife would take him to the cleaners, he gladly coughed up the money. "It's just business," Carmine would say. The guy had not been seen in that area of the city since.

As students shuffled into class, Carmine watched and waited for Kevin. He rolled through the door as the last bell rang for homeroom. He sat down behind Carmine and leaned over to talk to him. "Gizep, I'll have the cash for you in a few days. I have to move some money around. Can you give me the spread on the Flyers game tonight?"

"Well, the Flyers aren't favored to win, but you could get a long shot. How much do you want to put down?"

"I can get two hundred. If I pull through, I can pay you back the money I owe you."

"Or you could roll it over and bet on the Sixers game tomorrow night," mentioned Carmine casually. "Oh, by the way, did you see the paper this morning?"

"No," replied Kevin.

Carmine handed Kevin the front page of the *Philadelphia Inquirer*. In big letters across the top it read, "Manayunk Tormented by Brick Smasher." He read on to discover that the car he busted up belonged to one of the top aides of the mayor. Seeing how the aide had claimed the crime rate was declining in his last statement, he was eating his words. Now the pressure was coming down on the police to find the "barbaric hoodlum," as the aide was quoted in the article. "This person will be brought to swift and heavy justice," the aide went on. A large reward was being offered for any information regarding the incident.

Kevin gulped and gave a long sigh. "Good thing we didn't get caught. I made it down to the river and crashed by a log for a few hours. You made it home, didn't you?"

"Oh, yeah," Carmine smiled. "Those guys can't make it through the alleys, and I know those streets like the city slicker I am."

"Cool," said a relieved Kevin. "That was pretty close. I mean, I was drunk, too. I would have been put in the slammer for sure and my parents would have killed me."

"Well," Carmine interjected, "we can replay the memory as often as you like." He pulled out his cell phone and played back the incident to Kevin. The color drained from Kevin's face as he looked at Carmine with pleading eyes.

"Come on, man. You're going to erase that, aren't you?"

"Possibly," he said snidely. "What's it worth to you? If you get caught, you are a dead duck. Didn't you get nailed a few months ago for DUI? You would definitely be up the creek without a paddle."

Kevin instantly regretted the moment he had decided to accept Carmine's offer to play cards and score some easy cash. He had some wits about him, but not enough brain cells to realize that Carmine was nothing but pure trouble. Much to Kevin 's vexation, he was now in more debt than he bargained for when he picked up that first brick. The only plausible solution was to try and plead with Carmine to get rid of that film. One click and it could be sent directly to the police station. Kevin's miserable future was now at the mercy of a high school chump who would never amount to anything more than a repeat offender in the state correctional facility. The disconsolate Kevin had defeat written on his face like a blasphemy against the almighty. "You dirty son of a bitch," muttered Kevin. "I can't believe you would do that to me."

"Uh, hello, intoxicated brain-dead dummy. Remember I filmed you as I taunted you to throw the brick in the first place? You had the power to say no, but you insisted on proving your testosterone-driven ego and destroyed the car. It felt good, didn't it? That's what you said while the glass shattered all over the street. You're just contributing to the recidivism rate of the city's finest. Hey, I did you a favor; I took three hundred off your debt. I guess it'll have to go back up some more. Let's say fifteen hundred spread out over five weeks. That's three hundred a week. Seeing as you make pretty decent money at your job, three hundred a week is nothing."

Kevin let out a big sigh as his lips flapped together. He was stuck between giving up a majority of his paycheck for the next five weeks or going to jail, and one blemish on his record was enough for now. He reluctantly agreed to the deal. "What about the money I owe you now, Gizep?"

"Let's say that's your starting point. If you lose your bet, we start over and you still owe me fifteen hundred. It's just business. You're lucky I'm in a good mood, Kevin. I'm giving you a break and doing you a favor. I'm keeping your sorry butt out of jail. You should be thanking me."

Just then the bell rang, signaling the end of homeroom. Carmine gave Kevin a sarcastic smile and held up the camera phone. "I'll hold onto this until I receive the first payment. Then I will personally let you delete the film yourself."

"How do I know you won't make a copy and turn me in anyway?" asked Kevin in a shaky voice.

"Like I said, it's just business. I'm not looking to make any extra money on this. I'm just buying your freedom, buddy." With that, Carmine winked at Kevin and strolled down the hallway to his first-period English class.

∽ ∾

By lunchtime, Carmine was beginning to tire of his day. School was for entertainment purposes, and, of course, for seeking out potential opportunities to make a buck. He was just over 19 years old now, and there was no way he was going to graduate. His attendance record was mediocre at best, and he figured this would be his last year. Next fall would find him working the streets. He had delusions of grandeur about running his bookie business full time. At this point, Carmine was pulling in anywhere from $3,000 to $5,000 every two weeks. It was higher during football season, and the Superbowl was his biggest payday this year. Through his parlays alone he cleared $7,000. Needless to say, he hosted a small soiree for some friends who were entertained with lobster tails and petite filet mignons. His indulgence was nothing short of a king spending the money of his plebian servants who were there to provide him one thing: their money.

Carmine sat munching on a slice of pizza reminiscent of cardboard and washed it down with some lemon Gatorade. In front of him, he opened his log of bets. As he perused the pages, he wrote down amounts of money owed to him from the previous week. When he finished, he added the name of Kevin Sangiovesse and his new donation to the DiGuiseppe slush fund. He grinned devilishly and let out a diabolical chuckle as he thought how easy it was to make money on him. The words "pity" and "sympathy" had not been part of Carmine's vocabulary for a long time. After his parents were killed in a car accident and he was sent to live with his aunt and uncle, he quickly learned their love was not unconditional. He had everything he wanted, except for two people—now acting as his parents—who were willing to spend time with him and ask about his life. As the years went on, Carmine's heart grew more callous and his jaded perception of the world started to reward him financially. His charisma and knack for finding a person's weak spot flourished, and his bank account began to grow substantially. His motto was to look out for himself no matter what the circumstances. When someone was down on their luck, Carmine would reply with his usual, "It's just business." Nothing was personal to him. Either you owed him money or he owed you money; nothing more, nothing less.

That evening found Carmine sitting by the window of a local coffee shop and enjoying a mocha latte after his dinner. As he gazed at the passersby, he noticed two men standing attentively in front of Hikaru, the Japanese restaurant across the street. They stuck out because they were standing next to each other staring in his direction. A few minutes went by and they did not budge or talk to each other. Somewhere in the depths of Carmine's mind he pulled up a memory of two gruff Italian guys with thick South Philly accents. They were very knowledgeable in the realm of sports, betting, and how much money he was making. At that instant, Carmine's stomach shrunk and he felt a tightening of sphincter muscles in his buttocks. It was Vinnie and Jimmie from the supposed DeLuca Construction Company, one of the wealthiest construction companies in the city. "Construction" was not the first word that came to mind when he heard the name DeLuca. They handled all the bigger bets throughout the city and knew the local bookies. When money is made illegally, nothing goes unnoticed. Carmine sipped his coffee nervously because he knew they were waiting for him. He mustered up the nerve to go over and approach them. He was prepared to lie through his teeth and say it with authority and confidence to convince them. Popping a piece of gum in his mouth, he exited the coffee shop.

"Hey, I know you guys," Carmine said with a smile as he approached the two men. "You guys are from DeLuca Construction, right?" Carmine extended his hand and shook firmly with the two men who were decked out in black Armani suits and silk ties.

"Shut up, Gizep," said Jimmie. "We haven't seen you in a while, but the word is you're still running some bets around here. How ya making out on them? Collecting some money, are ya?"

"I don't run anything big anymore. Nothing over a thousand. I'm not ready to go too big. Then I would have to compete with you guys, and I'm not going there again."

"How am I supposed to know who can pay and who can't?"

"We hear you cleaned house on Superbowl Sunday," said Vinnie as he stepped closer. "Come on over here to the light so we can see you better. You look different from last time. Did you get a haircut or something?"

Vinnie could have cared less about Carmine's hair. He wanted to inspect him closely to see if he was telling the truth. Jimmie and Vinnie had cracked many a con in their day and continued to do so with brute strength and force. Their time spent in prison had hardened them even more and they knew when someone was lying.

"Say, Gizep," said Vinnie, "you remember that time when you was with your buddy and we had that big party over at DeLuca's house?" That was a great night, wasn't it? You won all those poker hands. What'd you win that night, like five Gs?"

"It was fifty-five hundred, to be exact," Carmine said with a smile, and started laughing. "Jimmie, I think you lost your shirt that night, literally. Did you ever win your Rolex back? Wow, what a night!"

"Yeah, well, those was good times, but now we need to talk about the new times. We heard from a reliable source you been making more and doing more than a grand on your bets. Where you getting the money to cover that?"

"I'm not," uttered Carmine. "I told you, I don't do anything over a thousand. I'm keeping a low profile. You know, I've been going back to school, and I'm trying to graduate this year and go on to college. I was just accepted to the University of Tennessee. Go Vols!"

"What you gonna major in, economics?" said Jimmie with a deep hearty laugh.

"You bet. I know numbers very well, as you know. They got a lot of hot chicks down there in the South. Maybe you guys could come down and visit. I'll take you to a game. The football team isn't so good right now, but their girls basketball team, now there's a winner. The coach has over four hundred wins. Believe it or not, some of the dudes around here are taking advantage of the action."

"Is that going through you?" Jimmie asked.

"Sometimes. I think some of the other guys have their sure-fires up in the Northeast section of the city."

"Why don't we take a little walk down the block here, Gizep," instructed Vinnie.

He led them off the street and down a path that ended under a bridge by the Schukyll. Carmine engaged in small talk to keep up appearances and never faltered in his voice or lines of bull. "So, how are things going in the construction business?" inquired Carmine.

"You know what," replied Vinnie, "things are going great. Mr. DeLuca is building stuff all over the city. You know, he's thinking about buying out some property down on the waterfront and putting up high-rise condos. He's betting against Trump right now, but blood is thicker than water and Mr. DeLuca's blood runs much thicker here than The Donald's. He can have Jersey and New York, but DeLuca owns Philly. So, Carmine, are you gonna tell us about the piece in your pocket or just let us wonder about it?"

"Come on, guys," shrugged Carmine. "I'm a businessman just like you. I need a little heat in case something comes up. It's like my American Express card; I don't leave home without it."

"Right," said Jimmie. "Let's cut the crap, Gizep. We know what you been up to. You been lying to us since we saw you sitting in the coffee shop. You're pretty good, but you ain't that good."

Jimmie locked Carmine's neck in the cradle of his arm. He was pushing 300 pounds, so there was no way Carmine could overpower him or reach for his gun. "Look, kid, we like you. You're funny and we hope you make it at the University of Tennessee, but for now, we're gonna tell DeLuca you ain't got nothing big for us."

With that, Jimmie grabbed Carmine's index finger and crushed it between his knuckles. The bones crunched audibly as Carmine's finger broke. He yelled out an obscenity as Jimmie let him go.

"What'd you do that for? Man, I been straight with you! Why would I lie to you? Look, I'm just trying to make some money to pay my way through college. I would never lie to you guys. Of course I was nervous tonight, for obvious reasons."

"Well, consider yourself lucky," said Vinnie. "You might be telling the truth, but we like our friends to remember who we are and what we can do. We find out you're doing anything over fifteen hundred and we're coming back with our pals Smith and Wesson. See that river there; we got a lot of friends who are now part of the permanent landscape at the bottom. The catfish are probably eating their eyeballs out. I'd hate to see a young man like yourself end up down there. Sorry about the finger, Gizep, it's just business. Hey, best of luck in college."

Jimmie and Vinnie turned around and walked away, leaving Carmine in agonizing pain, which did not show on his face, but he was reeling in agony inside. His contorted finger was bent like an upside-down L. Carmine took a deep breath, leaned down, and placed his hand on a rock. He took out his gun and slammed the bone back into the joint. He could wrap it up and hopefully it would heal itself. It was not the first time he had something broken and healed without a trip to the hospital. Getting away with his life and pulling off a bluff was enough success for one night. Placing his hand in his pocket, Carmine made his way back up to the streets of Manayunk and headed home for the evening.

෴   ෴

Vocabulary

Autonomic nervous system   A part of the peripheral nervous system that controls many organs and muscles in the body, which work in an involuntary manner.

Control questions   A question unrelated to the investigation, but which allows the subject to answer in a manner where he or she is unsure of the answer. It is then charted and used for comparison to answers that have been answered deceptively.

Electrodermal resistance   The ability of skin to conduct electricity.

Irrelevant questions   A question designed to elicit a truthful response used as a reference with relevant questions.

Physiology   A branch of biology that deals with the functions and activities of life or of living matter (as organs, tissues, or cells) and of the physical and chemical phenomena involved.

Pneumograph tubes   Two convoluted tubes that are placed around the chest during a polygraph procedure to measure respiration of the subject.

Polygraph technique   A technique that records physiological changes that can be used to determine if someone is truthful or deceptive.

Relevant questions   A question that is directly related to the investigation in which a subject is being questioned.

Forensic Information

Interrogation is one aspect of a crime scene investigator's job that requires tact and skill in reading the body language of potential suspects. When a suspect is apprehended at the scene of the crime, he or she may be questioned on the spot. Otherwise, it may happen at the police station or in the field.

There are several methods that detectives use to conduct interviews. If there is a possibility that the suspect may be dangerous, he or she should be checked or frisked for any kind of weapons and kept within view at all times. This is also a crucial time to pick up on body language of the suspect.

There are many cues that indicate a suspect is not telling the truth. Investigators may write these comments in a kind of shorthand on their paper during the interview. These cues may be as follows:

Break of gaze to right	R
Break of gaze to left	L
Clear throat	CT
Deep breath	DB
Delayed response	DR
Avoiding eye contact	EC
Grooming behavior	GB
Loud or obnoxious	!!!
Nervous laugh	☺
Fidgety	FY

Before the interview begins, the suspect should be able to answer the following questions without any hesitation:

- What is your name?
- What is your age?
- What is your occupation?
- What is your financial status?
- Do you have a criminal history or record?
- Are you related to the victim?

After those questions are established, there are several ways of interrogating a suspect:

- **Alibi:** Ask the suspect where he or she was during the crime.
- **Narrative:** Let the suspect tell his or her side of the story without interruption. The interviewer may use a tape recorder or a video camera.

- **Question and answer:** This seems to be the most common type of questioning. A question is asked, and the suspect answers while interviewer records or writes the response.

- **Sweet and sour (good cop, bad cop):** This method uses two kinds of interviewers. One is kind and interrogates in a calm matter. The other is harsh in demeanor and addresses the suspect in an aggressive tone.

- **Overheard conversation:** This method is used when two or more suspects are involved. When the second suspect is interviewed, the interviewer tells the suspect that his or her partner has already confessed to the crime.

- **Bluff:** This method is used to extract the truth from suspects. The interviewer places the suspect at the crime scene due to witness accounts that are made up. This method sometimes scares the suspect into making a confession.

(The previous two methods might be considered entrapment, depending on the attorney in the courtroom and the extent to which he or she will go to clear the client.)

There is no one method that works better than another. It depends on the suspect and how he or she responds to the method used.

As this information is gathered, it should be organized in some manner. Information taken from witnesses also plays a key role in establishing the validity of the suspect's deposition.

Sometimes as an investigation continues, a polygraph may be requested from the prosecution or defense. To many people, the polygraph is considered a test that tells when you are lying. There are no mechanical devices in existence that can automatically detect a lie. However, it is true that there is instrumentation available that can record physiological changes that may be a reliable basis for the diagnosis of truth or deception. This procedure is termed the polygraph technique.

The purpose of the polygraph is to make a permanent and simultaneous recording of a subject's respiratory rate; changes in electrodermal resistance, which is the ability of the skin to conduct electricity; and changes in blood pressure and heart rate. These physiological features are used because they reflect the activity within the autonomic nervous system. The three features are recorded as follows:

- Pneumograph tubes are fastened around the abdomen and chest of the subject for respiratory conditions.
- A standard blood pressure cuff is fastened around the arm.
- Two electrodes are affixed to the fingers of the subject and a minute trace of electric current is passed from one to the other to record the electrodermal resistance.

The examination is then conducted in a room that is free of noise and extraneous distractions. Preparation of the subject by the examiner requires that the examiner is provided with all case facts and investigative information. If the examination results are stipulated as evidence at the trial, the examiner will seek information from all sides as it pertains to the case.

Forensic Implications and Jurisprudence

There are three separate phases to the polygraph test:

1. A pre-test interview
2. The recording phase, which is charted
3. The diagnosis of truth or deception

Phase one is used to condition the subject and attempt to allay any fears or apprehensions that may hinder accurate chart recordings. Prior to the test, the subject is made aware, verbatim of the questions to be asked during the test. Surprise questions do not play a part in the procedure and would not be considered a reliable test.

Along with this discussion is the proper choice of control questions to be used. A control question is unrelated to the investigation, but it addresses behavior or motivation similar to the issue at hand. The most important factor here is that the control question must be one in which the subject will lie or be uncertain of his answer. The charted response will be similar to other relevant questions in which the subject is not telling the truth. For example, the subject will be asked what he or she had for dinner last Tuesday. If the subject is not sure, the charted response will be considered valid to measure against other questions asked.

Relevant questions are designed to specifically address the matter under investigation. These questions are critical to the ultimate inferences drawn from charted responses. Relevant questions must meet certain criteria to be considered relevant. For example, they should be worded in a simple language that is easily understood and direct. An improper way to ask a question would be, "Are you saying that the allegations against you have been fabricated by person X in order to have you fired from the company?" The proper way to rephrase the question is, "Did person X lie to have you fired?"

Another constituent for a relative question is to make sure that is not an assumptive one. For example, in a burglary case where the subject denies having brandished a gun in the air, it would be improper to ask the question, "Did you brandish the weapon to scare customers?" Questions of this nature could result in deceptive chart responses from a subject who is telling the truth. The examiner should not make assumptions when asking questions.

Irrelevant questions are those that have no direct bearing on the investigation, and are composed for the subject to tell the truth. These questions have an important function during the response charting process. First, they acclimate the subject to the testing procedure so normal physiological responses can be recorded before control and relevant questions are asked. These questions are also unique in that they do not elicit an emotional response that is noted in control and relevant questions. In some cases, a deceptive suspect will use irrelevant questions as an opportunity to answer incorrectly and skew the results of the test.

Phase two of the procedure is the charted response recordings. After an irrelevant question is asked, the examiner should wait approximately 10 seconds before the next question, and 5 to 10 seconds longer for relevant and control questions. Most tests are grouped in 10 questions. A minimum of three of these tests is considered the norm before a diagnosis can be made. Every situation is different and the examiner will determine what is appropriate to obtain the information for a proper diagnosis.

The last phase is the diagnosis phase. Generally, a subject who is telling the truth focuses emotional attention toward the control questions and away from the relevant questions. These responses are evident in the chart. On the other hand, someone who is not telling the truth focuses emotional attention toward relevant questions and away from control questions. In order for the criteria to be considered sufficient, the subject must be able to exhibit autonomic arousal and form perceptual distinctions between relevant and control questions. Some subjects do not exhibit either one of these, which can be due to medical or psychological problems, or because they are under the influence of alcohol or drugs. The diagnosis in those cases would be inconclusive.

For the most part, polygraph results are not admitted into court. They may be suggested by one side to strengthen the credibility of their client, but test findings do not meet the rules of admissibility. Approximately 65 percent of test findings are difficult to discriminate the truth from deception because they are very subtle in appearance. Although these interpretations are clear to the expert, they will not be obvious to a layperson in the courtroom. In approximately 25 percent of test findings, the results can easily be seen by the layperson. The remaining percentage is composed of findings that are inconclusive because of a subject's physiological, psychological, or other extraneous factors that prohibit successful results.

To qualify as a polygraph examiner, one must be intelligent with a college degree. The person must also be outgoing and personable, because test subjects come in an array of personalities that must be dealt with using discretion. Training consists of classroom instruction as well as an internship under a competent, well-trained examiner. The training should consist of exposure to a substantial volume of actual cases. Training in physiology, psychology, psychopathology, pharmacology, and law are essential to the makeup of the training examiner.

In order for an examiner to qualify as an expert witness, the examiner should (1) possess at least a bachelor's degree, (2) receive at least six months of training in actual case work under the direction of a competent examiner, (3) have three years of experience as a specialist in the polygraph field, (4) make results available for cross-examination, and (5) be able to provide a detailed explanation of the questioning hierarchy used and its importance in the interpretation of results.

6

Genetics and Pharmacokinetics

The sun was beaming on the morning rush-hour traffic speeding down the turnpike. Melinda and Fredric stood at the door of National Medical Labs where they waited to be buzzed in. Security was tight there, for the evidence in many high-profile cases was studied behind those doors. Melinda jumped as the driver of a passing truck slammed on his brakes, making a loud screeching sound.

"Whoa!" she yelled. "I don't think this building could be any closer to the road. Hassie, that scared the dickens out of me. I think I might have to change my underwear."

"That's more than I need to know right now, Monsternick," replied Fredric.

The door buzzed and clicked, signaling them to enter the building. The first two steps led them through a metal detector, which immediately sniffed out their guns and beeped like a tattletale. The receptionist looked up and smiled as she recognized the two. Melinda had been there many times, while Fredric was beginning to make the acquaintance of the lady behind the desk.

"Good morning, you two," she said sweetly, and fluttered a smile at Fredric. "How are we today?"

"Fantastic, Kristen. You remember Hassie—I mean Fredric—don't you?"

Kristen looked at Fredric and smiled shyly with a nod of her head. "How are you, Fredric?"

"Oh, no, please call me Hassie. That's what everyone calls me. I'm doing very well today. We're here to see Arthur Ying. Would you be able to give him a shout for us?"

"I sure can," she said with a promising smile. The amount of flirting was disgusting. Melinda kicked Fredric in the shin from behind the standing desk so Kristen could not see. Fredric chuckled in spite of himself and continued the small talk.

"So, you guys are always so busy here. How do you like working for one of the world's greatest crime labs?"

"I thoroughly enjoy it," she returned with a gleaming smile.

The phone beeped and Arthur gave the blessing for Melinda and her compadre to enter the lab. Fredric and Kristen exchanged a flirty goodbye.

The lab was the size of a warehouse converted into a crime-fighting repository. There were selected areas for specific fields of forensics. As Fredric and Melinda walked down the corridor, they passed various rooms such as trace evidence and the pharmacokinetics room. Fredric squinted one eye as he tried to decipher the hieroglyphics. "Monsternick, what is that?"

She laughed at him and slapped him on the back. "That, my friend, is where they study how chemicals are absorbed and metabolized in the body. They run all kinds of tests in there with urine, blood, tissue, muscle, fat, and whatever other part of your body can absorb a molecule. I took a class on it once and it blew me away how everything we eat, drink, and breathe can affect what happens inside our body. Pharmacokinetics is the study of all that stuff. Lots of mass spectrometry and spectroscopy happening."

"Right," said Fredric as he peeked through the window at the scientists in lab coats sitting at their computers, diligently studying readouts and punching in information for the computer to interpret. "I would imagine Art is going to explain the process of DNA and everything that comes along with it."

"You're darn tootin'," said Melinda. "He is one of the best. His experience runs far and wide. When you need a sample run or have a question about acquiring a DNA profile, Art is the man. All right, we have to go to the locker room first and suit up. This place is spic and span. The air is circulated with pure oxygen and dust free. The most minute of particles can mess up a sample. The DNA we're dealing with is mitochondrial DNA. It comes from the mitochondria inside of the cell and is located within the plasma. There are many mitochondria located throughout the cell, so it is more available than nuclear DNA, which resides in the nucleus of the cell. Mitochondrial DNA, or MTDNA, is useful in some cases because it has DNA only from the mother. When a baby is conceived and the egg is fertilized, the tail breaks off and the father's mitochondrial DNA is lost. Nuclear DNA has a fifty-fifty mix of mom and dad. Are you getting all of this down, Hassie?"

"Yeah," he said hesitantly. "I am scrupulously writing down mental notes in my head. Are we having a test later?"

"You betcha, so you better remember. Art will probably ask you a million questions, too. He loves to disseminate as much information as possible to everyone he talks to; you know, the scholarly type."

"Sure," boomed Fredric. "I love teachable moments. I'm learning way more than I could have ever imagined. I'm ready for my first true lesson in DNA. Let's do it!"

Arthur Ying met them in the locker room. "Hello, crime-busters, what's shakin', bacon?"

"DNA is shaking, Art," Melinda said with a healthy handshake. "This is my new sidekick, Hassie. He's expecting a complete education on DNA evidence and creating a profile that will speak volumes in the courtroom."

"Hi, Art," said Fredric, extending his hand for a shake. "I'm looking forward to a great day."

"Well, you shall have it," returned Art enthusiastically. "Let's get our gear on and head in. I'll show you some of the instruments we use and what happens in the process. We're going to be working with the PCR technique. Do you know what that stands for, Hassie?"

Fredric rolled his eyes upward searching for an answer. Unable to retrieve anything scientific, he replied with, "Pulmonary cardiac response?"

Art's eyes widened as he let out a gracious smile bearing his white pearly teeth. "You are absolutely, positively 100 percent for sure incorrect. You aren't even close, young Jedi. You have a lot to learn today. Come on Hassie, let's change your life by filling your head with crazy scientific terms, pronounce some names of chemical elements, and learn the difference between Fahrenheit and Celsius. Oh, yeah, we'll be giving the DNA a bath, too. I hope you brought your scrub brush."

After transforming into antibacterial clothing, they entered the lab. The room was in pristine condition. The floor was cleaner than a restaurant table, and the only noise to be heard was the spinning of the centrifuge whirling at an extraordinary 10,000 rpms.

Art started off with his basic introduction. Fredric listened eagerly as Art poised himself and took a deep breath.

"Young Hassie, DNA is the building block of life. Without it, you wouldn't exist, and we would look like blobs of slop. You are coded from the time you're conceived. From there, the DNA is the boss and tells everyone what to do. It consists of four basic proteins that are held together by covalent hydrogen bonds. The proteins are adenine, guanine, cytosine, and thymine. Each one of these is linked to a sugar group that is then bonded to a phosphate group. Together, they make what is called a nucleotide. They are countless because they are so microscopic and create the base pairs that bond together to form a double helix."

"I know the double helix," responded Fredric. "I remember that from my high school bio class. Now, correct me if I'm wrong, but the adenine can only pair with thymine, and guanine to cytosine."

"You are absolutely right, my friend," Art smiled. "They bond in sequences that make up a protein and in turn, they develop your physical features like hair color, eye color, and so on. All of the genetic codes are inventoried in those strands of DNA. Now most of everyone's DNA is the same. However, it is the so-called 'junk' DNA that separates us from the rest. In the sequence of DNA, there are repeats of codes throughout the strand. We call these short tandem repeats, or STRs. These make up the junk DNA that is matched to the sample, whether it be the victim or perp. Do you know where DNA is found within your body, Hassie?"

"I guess anything that contains cells, which is pretty much anything like blood, spit, semen, and urine."

"You got it, bro. Plus vomitus, feces, sweat, bone, tissue, hair roots, and even your teeth. All of them contain DNA. The first part of acquiring a sample requires the separation of DNA from the other elements of the cell, like the plasma and other organelles. We have to break the cells apart, so everything can leak out into the open. For this we place the sample in a buffered solution. Then we spin the test tube in a centrifuge to separate the DNA from the other stuff. What happens is the DNA has a higher molecular weight, so it precipitates out and is left at the bottom. Then we suck it up with a pipette and prepare for PCR. Do you remember what PCR stands for, Hassie?"

"You never told me, Art. I give up. What does it stand for?"

"It stands for polymerase chain reaction. It's a series of changing temperatures that forces the cell to undergo DNA division, creating millions of copies of DNA. Then there is enough DNA for the prosecution, the defense, and we always reserve a little bit as a referee sample."

"A what?" asked Fredric.

"Well, let's say that the prosecution tests the DNA and comes up with results we will call 'A.' The defense runs a test and comes up with results called 'B.' For some odd reason, both conclusions don't jive. Now the referee steps in. This sample is sent to a lab that has no interest in the case. It's tested, usually with accuracy, and then we find out who was trying to skew the results to benefit their case."

"That's a great idea, Art."

"Well, we do it whenever we can. We also urge the pathologists to take extra samples at autopsy so we have them. So, to continue, let's talk about the amplification process of PCR. The old process was one called RFLP. That stands for restricted fragment length polymorphism. This method requires at least a 50- to 500-nanogram sample for good results and takes up to two weeks to finish. RFLP requires a restriction enzyme digestion process in which the enzyme slides along the base pairs until it recognizes a specific pair. Then it will stop and digest or cut the DNA molecule at that particular area. This is called a restriction site. Sometimes it will cut at multiple areas, causing fragments of DNA. These fragments or 'digested DNA' are then loaded into an agarose gel and placed in an electrophoresis chamber. The chamber is then hooked up to an electrical source. DNA is negatively charged, so it works its way to the positive end of the chamber where the current is. The fragments are eventually transferred to a membrane made of nylon or similar material. Radioactive probes are applied to the membrane and the excess is washed away and the membrane is X-rayed. The result gives you a DNA pattern that can be analyzed or interpreted."

"Yowza!" exclaimed Fredric. "That sounds like a lot of time, money, and effort."

"Oh, yeah, it was painful. I tell you what, though, I had the opportunity to brush up on my solitaire skills on the computer. We weren't as busy back in the eighties so we had time to kill. You would be surprised what a room full of scientists will do to entertain themselves. We ended up pioneering some research in the field of forensic toxicology and made plans for a new instrument called the ICPMS. You know what that is, right?"

"Uh, no, not really. How about you, Monsternick, do you know what ICPMS stands for?"

"Why, yes, Hassie, I do," said Melinda as she pointed to an instrument through the window. "That instrument is called inductive coupled plasma mass spectrometry. It uses hot plasma to melt samples and detects heavy metals. It is highly sensitive and is excellent at identifying the elements of gunshot residue. How's that for an answer?"

Fredric nodded his head in approval. "I will never doubt you again, master."

"Yeah, right, Hassie. Remember, you'll have a test on this later."

"So, Hassie, let me explain the process of PCR," Art continued. "It only takes a couple of days to complete, and a .1 nanogram sample is needed. That means that a blood drop the size of a pinhead can be used. The sample can be somewhat degraded and still produce excellent results with much higher accuracy—one billion to one trillion!

"The premise of PCR works on the basis of temperature. We use a thermocycler because of its ability to heat and cool. First, the double helix has to be split. For this, the DNA is heated to approximately 94 degrees Celsius. This causes the splitting and exposes the bases. Do you know how to change degrees between Fahrenheit and Celsius, Hassie?"

"Yes. Well, I know that 32 degrees Fahrenheit is equal to 0 degrees Celsius. I think the conversion for Fahrenheit to Celsius is subtract 32, then multiply by 5 and divide by 9, and for Celsius to Fahrenheit, you multiply by 9, divide by 5, then add 32."

"Brilliant, Hassie. We always use Celsius because it's all related to the metric system. We scientists like to stay global, so we use the metric system. After a while it becomes second nature, just like your detective work. It comes with the job."

Art continued his explanation. "Moving along, the DNA is now split. Next, the temperature is lowered between 54 to 60 degrees Celsius where the annealing takes place. This is when the polymerase is allowed to bind to the single strand of DNA. The polymerase becomes the 'chief in charge,' so to speak. It translates the next base located on the chain. Lastly, the temperature is raised between 70 to 72 degrees Celsius, where the polymerase picks out the complementary DNA bases and adds on to the DNA chain. When the process is finished, the DNA is replicated the same as the original. The thermocycler runs through 32 cycles. So, the formula for PCR is 2 to the 32nd power, which equals about 4.3 billion copies of DNA."

"I'd say that's enough for everyone," interjected Fredric. "What if someone broke in here, stole the DNA, and planted it at a crime? They would be busted for sure."

"That hasn't happened to us yet. I am sure there are criminals out there wondering how they could do that. It's not easy being cheesy, Hassie. That's why crime doesn't pay."

"So once you have all of this DNA, how do you match it up with the unknown samples?"

"It's kind of similar to the RFLP method," Art explained. "There is a short core sequence that repeats. They are the short tandem repeats I talked about earlier. They are only three to seven base pairs in length. These STRs are statistically highly discriminating within the human population. So, have I totally put you into a tailspin, headwhack?"

"Kind of, Art," said Fredric shrugging his shoulders. "Some of this makes sense to me, while other parts are new and quite interesting. I'll have to borrow some books from my mentor over there."

"My library is always open, Hassie," Melinda said. "So, Art, can you run our samples for us now?"

"We're backlogged, as usual, but since this is a high-profile case, we can squeeze it in. Give me a few days and I'll be in touch. Let's hope we can match this up and at least confirm that the missing boy from Roxborough High has been found. Take care, guys, and I will talk to you in a few. Hassie, I hope you found today somewhat as an expansion of your horizons."

"Absolutely, Art. Thanks for everything and hopefully we'll be able to come visit again soon."

Three days later, Melinda and Fredric were sitting in the car enjoying Whoppers and milkshakes when Melinda's cell phone jingled the theme song from James Bond.

"I love that ring," she smiled as she flipped it open.

She answered with some uh-huhs and ended with a big thank you. "Well, Art says it's a match. The body in Fairmount Park was none other than Phillipe Mindrago. What a shame, but the parents will be able to get some kind of closure. We'll have to head over there later to inform them. Hassie, it's one of those things you never get used to. I have met death many times and told a lot of sad souls about their loved ones. It's the most difficult part of this job, and it never gets easier. Kids get you in the gut. This kid was on his way to stardom down at Virginia Tech. Did I ever tell you that my cousin is the quarterbacks coach there?"

"No, I didn't know you liked football."

"I don't," shrugged Melinda. "He's just my cousin and I don't see him too often. He is one heck of a coach, though. They finish in the top ten almost every year. So Hassie, what is your SWAG?"

"My what?" questioned Fredric with a high-pitched voice.

"You know, some wild anonymous guess."

"Oh. Well, part of me wants to say serial killer because we have two male Caucasians, both of whom are high school students at the same school. A serial killer would profile similar people. However, given the Drater Club book and all of this gambling, I think these guys bit off more than they could chew and couldn't pay up. Do you think the mob might be involved?"

"It's hard to say," said Melinda as she slurped the last of her strawberry milkshake. "If these guys were betting through some chump, it might explain the execution-style gunshot, but I don't think gangs really run numbers that much. Drugs, yes, but I don't think numbers. It could be mob-related, but I don't think those guys bother with high school knuckleheads. I really can't say I have any strong convictions at this point. What we really need to do is talk with our friend Carmine DiGuiseppe. We know he's probably the bookie at the school and running the bets. It sounds like he is more of a scam artist, not a murderer. But maybe as he progresses, he's becoming more audacious and willing to step up to higher crimes. We have to be in court for the next two days, so let's set up some Carmine time for this weekend."

∽ ∾

Vocabulary

Adenine   One of the four nitrogen-containing molecules present in DNA. Designated by the letter A.

Alleles   One of a series of alternative forms of a gene at a specific locus in a genome.

Amino acids   Nitrogen-containing compounds that are the building blocks of proteins.

Annealing   The pairing of a mixture of two complementary single-stranded nucleic acids to form double-stranded (duplex) nucleic acids.

Chromosomes   Structure of DNA and associated proteins that contains the hereditary material within the cell. Genes are organized in a linear arrangement in the chromosome.

CODIS   Combined DNA Index System.

Cytosine   One of the four nitrogen-containing molecules present in DNA. Designated by the letter C.

DNA   Deoxyribonucleic acid, the stable double-stranded helical molecule that makes up chromosomes.

DNA fingerprinting   A process that uses fragments of DNA to identify the unique genetic makeup of an individual.

Dot blotting   A DNA analysis system where sample DNA is directly pipetted onto a membrane, as opposed to the Southern blot procedure of enzymatic digestion, electrophoresis, and Southern transfer.

Electrophoresis   The process of separating charged molecules in a porous medium such as agarose, by the application of an electric field.

Genes   Unit of heredity that is a region of DNA containing the blueprint for specific RNA formation or regulation of the formation.

Guanine   One of the four nitrogen-containing molecules present in DNA. Designated by the letter G.

Hybridization   Process of complementary base pairing between two single strands of DNA.

Locus   A specific position on a chromosome.

Mitochondrial DNA   The DNA that is located in the mitochondria of the cell. It contains maternal DNA only.

Nuclear DNA   The DNA that is located in the nucleus of the cell. It is a 50-50 mix of your mother and father.

Nucleotide   A combination of a base with a sugar and a phosphate group.

PCR   Polymerase chain reaction. Uses temperature to amplify DNA for testing purposes.

Pharmacokinetics   The study of the metabolism and action of drugs with particular emphasis on the time required for absorption, duration of action, distribution in the body, and excretion.

Polymer   Large organic molecule formed by combining many smaller molecules (monomers) in a regular pattern.

Proteins   The active molecules in all cells. Proteins control biochemical reactions and determine the physical structure of organisms.

Restriction enzyme   Enzymes that cleave double-stranded DNA at specific base recognition sites. The sequences at restriction sites are usually for each enzyme.

RFLP   Restricted fragment length polymorphism. An older form of DNA amplification.

Southern blotting   The procedure for transferring denatured DNA fragments from an agarose gel to a nylon membrane where they can be hybridized with a complementary DNA probe.

Thymine   One of the four nitrogen-containing molecules present in DNA. Designated by the letter T.

Background Information

Because DNA is such a huge field of information, I'll provide a general background here. Please see Appendixes A and B for sources of more in-depth information.

Deoxyribonucleic acid, or DNA, is responsible for understanding the underlying concepts of inheritance. Since its discovery, scientists have been working diligently to unravel the mystery it contains and pioneer the field of genetics. In 1985, a man named Alec Jeffreys and his colleagues at Leicester University in England discovered a way to isolate and interpret DNA, hence the birth of DNA fingerprinting. The methods have developed in leaps and bounds; today scientists can duplicate DNA from samples smaller than a pinhead, creating billions of copies of DNA.

DNA is a polymer, which means that it is a large molecule made by linking together a series of repeating units. Each of these series is called a nucleotide, which is composed of a sugar molecule, a phosphorous-containing group, and a nitrogen-containing molecule called a base. There is no limit to the length of a DNA strand, and some may be composed of long chains containing millions of bases. These strands coil into a double helix, giving the shape of the DNA molecule. There are four bases that constitute the double helix: adenine, thymine, guanine, and cytosine. Adenine always joins with thymine and guanine with cytosine. Each base is connected to a sugar group, which is connected to a phosphorous group, which connects to another sugar group connected to another base.

Heredity is passed on by means of microscopic units called genes. It is the basic unit that determines and controls the development of specific characteristics in a new individual. They are responsible for the growth of virtually every body structure.

The genes are positioned on chromosomes, which are threadlike bodies that are housed in the nucleus of every cell. All human cells contain 46 chromosomes that are broken down into 23 pairs. The only exception to this is the sperm and egg, which contain 23 unmated chromosomes.

While chromosomes come together in pairs, so do the genes they bear. The position of place that a gene occupies on a chromosome is known as its locus. The genes are in a similar position to that of the matching pair. Therefore, the gene for hair color on the mother's chromosome is aligned with a gene for hair color from the father's side. Alleles are chromosome pairs that influence a given characteristic and are aligned with one another.

Proteins are complex molecules that are produced and controlled within the DNA. These are created by linking together a combination of amino acids. There are thousands of proteins that exist, but all of them are derived from a combination of 20 known amino acids. The sequence of amino acids in a protein discerns the shape and function of the protein. For example, the protein hemoglobin is found in red blood cells. It is responsible for carrying oxygen to the cells in our body and also removes carbon dioxide from cells. By knowing the string of amino acids that comprise hemoglobin, problems with red blood cells, such as sickle cell anemia, can be spotted by looking at the substitution of one amino acid that creates the disease.

Each amino acid is coded by three bases. For example, phenylalanine is coded A-A-A. Alanine is coded C-G-T. As these proteins come

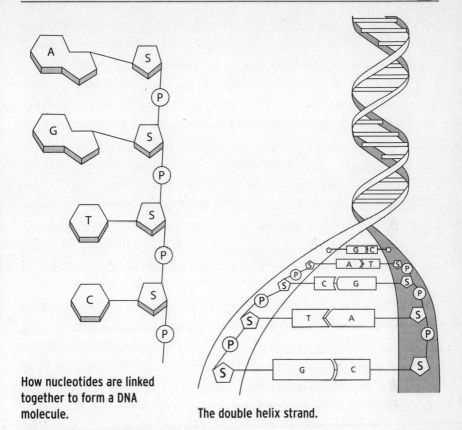

How nucleotides are linked together to form a DNA molecule.

The double helix strand.

together, they create the strands within the protein. It is these strands that later become short tandem repeats or STRs that scientists reproduce to match with a sample.

The key to understanding DNA typing is the premise that within the population of the world, there are numerous possibilities in which a particular base of letters can repeat themselves on a strand of DNA. The possibility increases when one examines two chromosomes, which contain different lengths of repeating sequences. Once these are amplified, they can be compared for identification.

DNA was first analyzed using the RFLP (restricted fragment length polymorphism) method. A 50- to 500-ng sample (about the size of a dime) is needed to perform the test and the condition of the DNA has to be heavy in molecular weight and intact. It also uses six areas of comparison that give up to 1 in 1 hundred million to 1 in 1 billion odds. RFLP requires a restriction enzyme digestion process in which the enzyme slides along the base pairs until it recognizes a specific pair. Then it stops and digests or cuts the DNA molecule at that particular area. This is called a restriction site. Sometimes it cuts at multiple areas, causing fragments of DNA. These fragments or "digested DNA" are then loaded into an agarose gel and placed in an electrophoresis chamber. The chamber is hooked up to an electrical source. DNA is negatively charged, so it gravitates to the positive end of the chamber where the current is. The smaller fragments move at a faster rate than larger fragments.

Once electrophoresis is completed, the fragments are chemically treated so the double helix separates. The fragments are then transferred to a nylon membrane in a process called Southern blotting, named after its inventor, Edward Southern. The sheet is then treated with radioactive probes containing a base sequence complementary to the RFLPs being identified. X-ray film is placed next to the membrane to detect the radioactive pattern; once it is developed, the DNA fragments that combined with the radioactive probes appear. The result gives you a DNA pattern that can then be analyzed or interpreted.

Bands appear that can be compared to victims, suspects, and crime scene samples. The number of visible bands is a direct result of how many locus probes are used. A single-locus probe results in one or two bands, while a multi-locus probe can have many bands with varied intensity (darker or lighter). The darker the band, the better the match.

By using a single-locus probe, results show sequences that are identical or similar to that of the selected probe. When using this method, a higher sample is needed, because more DNA is being analyzed. A multi-locus probe is very effective because the chances of two individuals randomly matching all band positions are very rare.

The latest method used in the lab is PCR, which stands for polymerase chain reaction. It is much quicker and easier than RFLP, taking a much shorter time to complete. Only .1 to 1ng is needed for testing. That means that a blood drop the size of a pinhead can be used. The sample

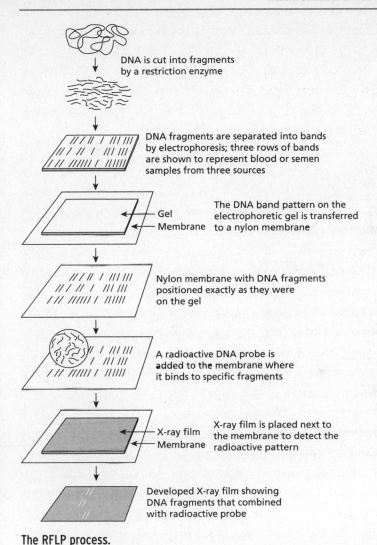

DNA is cut into fragments
by a restriction enzyme

DNA fragments are separated into bands
by electrophoresis; three rows of bands
are shown to represent blood or semen
samples from three sources

Gel
Membrane

The DNA band pattern on the
electrophoretic gel is transferred
to a nylon membrane

Nylon membrane with DNA fragments
positioned exactly as they were
on the gel

A radioactive DNA probe is
added to the membrane where
it binds to specific fragments

X-ray film
Membrane

X-ray film is placed next to
the membrane to detect the
radioactive pattern

Developed X-ray film showing
DNA fragments that combined
with radioactive probe

The RFLP process.

can be somewhat degraded and still produce excellent results. The accuracy is much higher, too—1 in 1 billion to 1 in 1 trillion! An enzyme called the DNA polymerase is the most important factor of PCR, because it can be directed to synthesize a specific region of DNA.

The premise of PCR works on the basis of temperature. Scientists use a thermocycler because of its ability to heat and cool. First, the double helix has to be split. For this, the DNA is heated to approximately 94

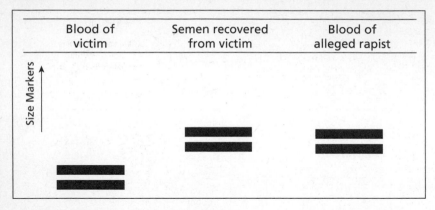

DNA/RFLP pattern represented by bands.

degrees Celsius. This causes the splitting and exposes the bases. Next, primers are added to separate the strands and allow the strands to combine. In this phase, the temperature is lowered to 54-60 degrees Celsius. This is also referred to as the annealing phase. The polymerase becomes the "chief in charge," so to speak. It translates the next base located on the chain. Lastly, the DNA polymerase is added along with a mixture of free nucleotides composed of adenine, guanine, thymine, and cytosine. The temperature is raised to 70-72 degrees where the enzyme directs the rebuilding of the DNA molecule, adding the appropriate bases one at a time, creating two complete pairs of double-stranded DNA segments. When the process is finished, the DNA is replicated the same as the original. The thermocycler runs through 32 cycles. So, the formula for PCR is 2 to the 32nd power, which equals about 4.3 billion copies of DNA. Now there is enough DNA for everyone to test. Defense lawyers will have some for their scientists to test as well as anyone else who needs to testify in the case.

The results are interpreted differently compared to the RFLP method. A dot blot system is used in which dots appear in specified areas to display a match. In the lab, this is known as sequence specific oligonucleotide (SSO) or allele specific oligonucleotide (ASO). This involves a specific probe that is a short oligonucleotide that ranges from 15 to 30 nucleotides in length. This sequence is identical to the target allele.

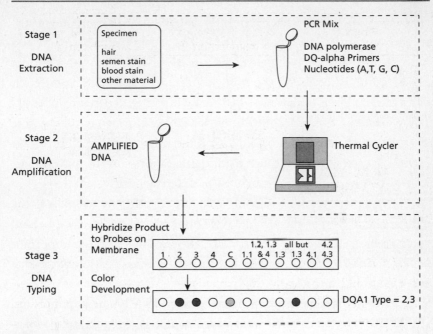

PCR process with dot blots.

Without going into too many in-depth terms that would require more than several pages to explain, think of it as a series of dots with numbers assigned to each location. After the dot blotting is complete, certain dots will appear that match with victim, suspect, and crime sample.

The use of specific primers and sequence-specific oligonucleotide probes, as described above, is one of the best-developed procedures used in the forensic lab. It has been subjected to extensive testing and validation has been granted by professional organizations because of the discriminatory power and the simplicity to use it.

Forensic Implications and Jurisprudence

There are two sources of DNA. One is nuclear DNA, which is found in the nucleus for the cell. This material determines what kind of physical characteristics people will have. It is a half-and-half mixture of the mother and father. It is found in the nucleus of every cell in the body and remains the same forever. The other source of DNA is mitochondrial DNA, which is located in the mitochondria of a cell. This is different from nuclear DNA in that it is maternally inherited. Therefore, paternal DNA does not appear in mitochondrial DNA. During conception, the tail falls off the sperm, therefore losing the mitochondrial DNA of the man. It is passed on from generation to generation, so your mitochondrial DNA is the same as your furthest back female relative

DNA plays its most valuable role in the courts by identifying the perpetrator of a crime through analysis of biological evidence left behind at a crime scene or on an object used by the criminal. Another important consideration is evidence found on a suspect or in possession of the suspect that links him or her to the crime.

Paternity testing is another area where DNA typing is important. Aborted fetuses, conceived as a result of a rape, have been linked to defendants through this method. When body parts are found or dismemberment has occurred, investigators have used typing to identify murder victims when no other sources are available. Serial crimes also employ DNA typing where a common perpetrator is involved or to negate cases thought to be related. Lastly, DNA typing can be used to exonerate those who have been falsely convicted or have been sitting in prison, because DNA testing was not available during their trial. It serves many purposes.

A system is being implemented to build a database of offenders based on their DNA. This system is called CODIS, which stands for Combined DNA Index System. It works the same way that AFIS (Automated Fingerprint Identification System) does.

DNA can be found in:

- Blood
- Hair shafts and follicles
- Dandruff
- Saliva
- Semen

- Skin

- Fingernails

- Human waste

- Bones

- Teeth

- Cigarette butts

- Licked postage stamps

- Chewed gum

- Toothbrush, hairbrush, or other personal items

When collecting blood for DNA samples, the same protocol as discussed in Chapter 2 is followed. Items containing DNA should be packaged in paper evidence bags, pillboxes, or paper using a druggist's fold. No DNA evidence should be placed in plastic due to the possibility of bacterial or moisture development that can destroy or degrade the evidence. Any potential DNA evidence cannot be compromised by mishandling. Proper handling includes documenting the evidence with pictures, video, drawings, and/or narratives; correctly labeling the package; and wearing gloves and other appropriate clothing.

When using RFLP, misinterpretation may occur because additional bands are visible. This may result from several factors, such as:

- **Sample cross-contamination.** This may occur at the time of collection or at the laboratory. Additional bands may be created that can skew the interpretation process. Another concern in cross-contamination is lateral movement of DNA. Sometimes during the electrophoresis process, the DNA can move into another lane of the gel plate. This can easily be remedied by leaving every other lane empty in the electrophoresis chamber.

- **Mixed stains.** This is different from cross-contamination in that DNA samples from separate sources may be mixed together when the sample is obtained. This poses an interpretation difficulty because the male and female lanes may show extra bands belonging to the other sex. This mishap can be avoided by running known samples from the contributors of each sex.

- **Partial digestion.** Restriction enzymes cut or cleave the DNA molecule at specific sites, creating restriction fragments. When digestion is incomplete, the molecule is not cut at the appropriate restriction site and only portions of the site are digested. The extra bands appear because the fragment lengths are longer and have a higher molecular weight than the properly cleaved DNA. If this is the case, the bands appear at the upper end of the gel plate, because they do not migrate far because of their size.

- **Star activity.** This is the other side of partial digestion in which the restriction enzyme cuts or cleaves at too many restriction sites, creating a possible extra band, which can lead to misinterpretation. This may result by a poor choice of enzymes that do not have a high degree of specificity for the sample.

- **Contamination.** Bacteria plays a significant role in contaminating evidence. Several biological substances may have been mixed in, other organic materials may have contaminated fluids and tissues, and the environment itself may contribute as well. More specific probes are being used so bacterial contamination becomes a non issue. As research continues, scientists are finding ways to create single-locus probes that are not deterred by bacterial and microbial contaminants.

Misinterpretation of PCR outcomes may result as the previously mentioned factors related to RFLP. This is because a number of the same procedures used in RFLP are applicable to PCR as well. In addition, factors like qualitative and quantitative fidelity are factors that may cause difficulty for interpreting results.

- **Qualitative fidelity.** If there is a low specificity during the primer hybridization process, unintended products may appear, which may be further amplified, creating more changed sequences. These may be attributed to the length and sequence of the primers as well as the temperature of the annealing process. Alignment errors may also contribute the qualitative fidelity. When there is a misalignment with the primer and template strand, sequences that contain tandem repeat sequences may cause problems. This may result in a heterogeneous collection of fragments, creating difficulty for the person interpreting the results.

- **Quantitative fidelity.** While amplifying the DNA, it is important to determine why some alleles amplify more efficiently than others. It can be affected by the length of the targeted sequence. Shorter sequences tend to amplify more efficiently, resulting in a preference for shorter sequences. It may also be caused by amplification of sequences that have significant differences in guanine/cytosine content. This instance requires higher temperatures for denaturation of strands. If the temperature is not high enough, the guanine/cytosine contents may clap together. To ensure full strand dissociation, the scientist should consider optimal denaturation duration and temperature. Another factor that can affect amplification is base pair mismatch. Where the mismatch is determines the area that is affected, whether it is at the end of the primer or the interior.

Several organizations have been working together to validate institutions and laboratories to keep them accredited in the name of justice. The recommendations for quality assurance contained in some of these guidelines are endorsed by the American Society of Crime Laboratory Directors (ASCLD).

The court allows experts to testify about results of tests performed by other technicians under the expert's supervision when the testing was done. It is generally accepted in the scientific community that a scientist can view results prepared by another, determine their quality, and determine whether there is a match. These experts can also testify to the laboratory's protocol and procedures for analysis of the evidence. This testimony is a lot more effective if visual aids are used and the expert takes time to explain DNA typing in layman's terms.

The responsibility of the DNA analyst is to perform DNA profiling on biological samples, interpret test results, provide expert witness testimony, perform internal validation studies, and participate in training classes updating new theory and practice. Those interested in becoming a DNA analyst need a bachelor's or master's degree in biology, chemistry, or forensic science. A minimum of six months of forensic DNA casework experience is necessary to qualify for the forensic analyst position. From there, three additional years of forensic experience must be completed to become a technical leader. Along with this training, classes must be attended in the fields of biochemistry, genetics, and molecular biology.

7

The Manayunk Punk

The bell clanged as students squeakily shuffled to their third-period class. Melinda and Fredric sat in the office waiting for Principal Monk, who was currently disciplining a student for pulling the fire alarm. It would not have been so bad, but the rain was coming down in droves. Students and faculty alike were drenched with the unexpected evacuation of the building. The secretary was wiping off her fogged glasses from the change in temperature. "Hello, detectives, Principal Monk will be with you momentarily. He's taking care of the putz who pulled the fire alarm. Let me buzz him, you might be able to talk some sense into this kid."

Principal Monk opened his door and invited them in. Across from his desk sat a young lady in a white shirt way too short for her stature. Seeing she had just been in the rain did not help either, for following the prescribed rules of fashion, it was hard not to notice the obvious lack of undergarments. Principal Monk walked back around to his desk and sat down. The beads of water glistened on his fuzzy black hair. He clasped his hands and started his tirade. "So, as I was saying, I could have you cited for falsely pulling a fire alarm and endangering the students by sending them outside in a thunderstorm where lightning was striking everywhere. I also see you have no recollection of the dress code, because you look like a walking billboard for a beer commercial, not to mention that everyone will probably be sick because it's damp and cold and everyone has to walk around like that for the rest of the day. I hope you're happy, Elise, because you just bought yourself ten days out of school! I know your mother will not be happy when I call her. Are you understanding what I'm saying to you?"

With that he pointed his finger that was now shaking from his heightened emotions. Elise jumped out of her chair and exclaimed, "My dad has a watch just like that! Where did you get it? I bought his at Lord and Taylor at the King of Prussia Mall. Where'd you get yours?"

Amazed, Principal Monk looked over at Melinda and Fredric who were just as shocked. The visible veins started pulsating in Monk's head. "What's wrong with you? Are you on something? Didn't you listen to a word I just said? Get out of here and go to the nurse! I want you checked out immediately! I..."

"Wait," said Melinda. "Let me get a squad car over here. We can take her to a hospital and get a blood sample for testing. We'll also write a citation for setting off the fire alarm."

Elise was directed to another room while awaiting her chariot to the hospital. Back in the office, Principal Monk sat down in his chair and shook his head. "I don't know what the hell is wrong with these kids. I've had some whackers, but never like that."

Just then the nurse knocked lightly on the door and stuck in her head. "Uh, excuse me, Principal Monk, but Elise didn't take her meds today."

"What meds?" he asked with a furrowed brow.

"The ones that keep her from going bonkers. She has a severe case of bi-polar, with a nonverbal communication disorder that's only accentuated by loud noises such as fire alarms. She's not on anything. You might want to apologize to her. Also, it says in her individual education plan that when she doesn't take her meds, she's exempt from all discipline problems or episodes that might otherwise implicate her."

"Right," he said with a huff. "I'll be there in a minute."

She closed the door and left. "Detective Monsternick, I can't tell you how many excuses these kids come up with to blame their behavior. Their parents take them to doctors and have them tested for everything under the sun, and when they can't find anything wrong with them, they go to some big-wig specialist in the city and concoct a rare learning disability that causes bad behavior. People say I'm insane for doing this job. I don't suffer from insanity; I enjoy every minute of it. Boy, if the board of correction returned, look out."

Melinda and Fredric were not sure how to respond to that one, so they both smiled and Melinda politely asked if they could have a chat with Carmine DiGuiseppe. Principal Monk buzzed the secretary, who returned a minute later to report that Carmine was on "vacation" today and not taking any calls right now.

"Thanks," replied Melinda. "We'll be in touch."

As they walked back out to the car, the rain had slackened and the sun was beginning to peek its head through the midday clouds. Once again Carmine eluded the grip of justice just as he had done before. They sat in the car watching the passing clouds. "Hassie, I think we need a stakeout tonight. I'll call the station and let them know our plans. We'll start at his house and go from there. Aren't you glad you don't have to be in charge of a thousand students plus a faculty?"

"I could never do that job," laughed Fredric. "I have a hard enough time dealing with the pimply punks on the streets. Not my cup of tea."

"Well, buddy, over time you'll begin to realize that some people are alive only because it's illegal to kill them." They both laughed and went to lunch.

⤸   ⤸

Later that afternoon, the sun was shining bright on the streets of Manayunk. Across the river, the traffic was slowly backing up like an overused toilet at a frat party. Melinda and Fredric sat contentedly in her little Honda Civic. She called it her college car, because that was what she thought college kids drove. Fredric was munching on some red pistachios. His red-stained hands worked quickly as he shelled the nuts and dropped them onto a napkin between his legs. His rosy lips resembled the early stages of Bozo the Clown putting on makeup for the circus. Melinda couldn't resist the opportunity for a few shots. "Why don't you just take some of my lipstick and put it on like a drag queen?"

"I love the red ones, Monsternick," he replied.

"What would you do if DiGuiseppe came out of his house right now and you had to bust him? I would laugh in your face if you asked me to put my hands up. Would you grab him with your goopy-coated hands and wrestle him to the ground? I'm curious."

"I would just wipe it off real fast and go after him."

"Right. I remember one time I was on the streets looking for unsuspecting hoods up to no good. I turned the corner and saw this guy taking a roll of cash from a girl, who had a baggie full of something. I immediately yelled for them to put their hands up. The girl took off and the guy raised his hands up above his head and his pants fell down around his ankles. At this point he didn't know whether to run, pull his pants up, or stand there dangling in the wind; he didn't have any underwear on. I asked him how it felt to be caught with his pants down. He didn't find it funny."

As the hours passed with recollections and anecdotes from their younger years, dinnertime was approaching. Fredric's stomach rumbled first, triggering another one from Melinda. "We haven't seen squat," she said. "I was hoping he would be home or at least dolling himself up to go out. Let me go knock on his door and see if anyone is home. If not, we go eat."

After the doorbell rang, an unkempt lady with scraggly black curly hair answered the door. The unpleasant smell of stale cigarettes and booze permeated the air as she greeted Detective Monsternick with a, "Whatta you want?"

"Hi, ma'am," said Melinda with a jovial smile. "I was wondering if Carmine was home. I was told he could help me in picking the right teams for the games this upcoming week."

"Well, he ain't here. He don't show up much around here. Try down on Main Street. Sometimes he hangs out down there with his buddies."

The lady squinted as Melinda moved to the side, blinding the silly sot with the sunlight. She could tell the lady was soused from the stench of alcohol on her breath and Lord-knows-what-else flowing through her veins. "Well, thank you for your time. You enjoy the rest of your day."

Melinda walked back over to the car, crossing her eyes and sticking out her tongue at Fredric, who was laughing because he'd watched the entire conversation transpire.

"What a waste," chuckled Melinda. "She should go shave her mustache off and cut herself to let the blood drain out of her eyeballs. You would have liked her choice of perfume, Hassie; the granola head special."

"You mean petroleum oil and stinky unshaved armpits? No, thanks. Bad enough I have to arrest them. What did she say?"

"She said we should go eat dinner and check back later. She needs a chance to shower up and get ready for you."

Fredric blew out his cheeks and he feigned puking from repulsion. "Let's eat."

Manayunk had become one of the dining hubs of the city. There was a plethora of fine restaurants all with their own eclectic dishes. Smells of delectable breads filled the air as they entered Le Bus, the restaurant famous for supplying gourmet bread to the county of Philadelphia. The sun was close to falling asleep for the night, but the air was warm. "Hassie, let's eat outside. Maybe we'll get lucky and we can watch the sights of Manayunk. I love watching people."

"Whatever floats your boat, Monsternick," he replied.

"You have to order the Chilean sea bass. It's the finest I have ever eaten."

"I'm not a big fan of fish. How about meats? I can always chow down on some beef."

"Well," said Melinda, "if memory serves me correctly, I think tonight is prime rib night. It comes smothered with au jus and the fat melts in your mouth. Too bad we can't drink. It goes great with an oaky cabernet."

"I like beer better. So what do you think about this DiGuiseppe guy?" questioned Fredric.

"It's hard to say. A lot of times these chumps start out small. First, they realize they're smart and can outwit most of the schmoes walking the streets. The more they swindle and hustle people, the more brazen they become. During one of my criminal justice classes, I spent some time in a prison for the criminally insane. I had the opportunity to speak with a guy who was somewhat of a serial killer. He killed three women, then his mother, and then was caught, which is enough for me to call him a serial killer. This guy was a freak. This is kind of funny, but not really. After he killed his mother, he cut out her voice box and put it in the garbage disposal because she was always yelling at him. He thought he could shut her up forever. You know what? When he turned on the garbage disposal, it shot her larynx back at him and hit him in the face. She was yelling at him from the dead. As a youngster, he used to kill things and then set them on fire. For him that was as exciting as you watching the national championship of college football. Anyway, he started stealing pets from people's yards and doing the same thing. Next thing you know, he was doing it to helpless college students at the University of Drexel. There seems to be three common denominators among kids who turn into serial killers: bed wetting, fire setting, and cruelty to animals. Granted, it's not with all serial killers, but there is a pattern. If you ever have the chance to take a class on forensic psyche, I highly recommend it."

"So how would you profile DiGuiseppe?"

"I think he is definitely capable of murder, as all of us are. We just know right from wrong and don't act on those thoughts. He feels comfortable enough taking advantage of people and is not afraid of consequences at school. Chances are he's not afraid of a little jail time, either. He doesn't care about his education, can lie like a rug, and can outsmart the savviest of street people. He probably thinks the world owes him and he hates everyone. A pure misanthrope. I would imagine he knows where his limitations are and keeps under the radar. To venture another guess, I bet he likes to flash his money around and wear fancy clothes and expensive jewelry. His glitz and glamour are just another part of his charismatic personality that sucks people in."

"Are you ready to order yet?" asked the waitress, who was focusing her attention on Fredric.

Melinda started first. "I'll have the sea bass and champagne vinaigrette on the salad."

"And for you, sir?" she inquired attentively.

"I would love to eat your prime rib."

She looked at him incredulously. "Excuse me?"

"I'm sorry. I would like the prime rib cooked rare. Must have been a Freudian slip."

"Would you like anything to drink with that?"

"Just a Diet Coke. I'm on duty."

"He has to work the graveyard shift tonight, kiddo, but thanks for asking," interjected Melinda. "Fredric, you can come back later."

All of them laughed and the waitress trotted back inside. The traffic on the street was beginning to congregate as the Friday-night happy hour started. The youngsters were souped up in their finest duds with one mission in mind: a possible companion for the evening. Melinda and Fredric sat anxiously waiting for their dinner to arrive. The bread disappeared five minutes after it was delivered to the table. Melinda lifted her glass to suck out an ice cube and stopped as her eyes wandered across the street. "Erw ee iz."

"What?" asked Fredric.

She spit the ice cube back in her glass. "There he is across the street at three o'clock."

Fredric turned and glanced across the street. A half a block away, Carmine DiGuiseppe was strutting down the sidewalk like John Travolta in the movie *Saturday Night Fever*. He had on a black silk shirt that glistened in the neon lights from the bar where he now stopped. The gold chains were visible from across the street. "What should we do?" asked Fredric.

"Here's your dinner," said the waitress.

"Great, you can just put them down here," gestured Melinda. The waitress was confused how such a previously congenial young man could be so cold. Fredric didn't even look at her. His eyes were set on the prize of the evening, Carmine.

"Hassie, not yet. Let's see what happens. Get some cash ready and put it on the table in case we have to leave before our dinner is over."

Across the street, Carmine stopped in front of a store where he nodded to a young gentleman who was standing there. It was obviously a meeting that had been previously established. The two talked for several minutes while Melinda and Fredric took turns stuffing their faces with food and watching him. Then Carmine gave a tap of the knuckles, turned the corner, and disappeared. "All right, let's go," signaled Melinda.

They crossed the busy street and turned the corner to see Carmine walking away, making a call on his cellphone. "Perfect," said Melinda. "He has a distraction while we tail him."

Carmine stopped momentarily as he tried to find a signal. Melinda and Fredric pretended to window shop at a jewelry store with diamond rings on display. Carmine observed them standing there, but did not see them as a threat. His signal came back and he continued walking. As they tried to conceal themselves the best they could, Carmine kept glancing back at them. Somehow they were becoming suspicious in the back of Carmine's mind.

"He's on to us," muttered Melinda. "I'm going to yell up to him and pretend like I know him. Then we can get him to stop." Melinda cleared her throat and cupped her hands around her mouth. "Yo, Carmine!" She yelled up the street. "Yo, Carmine, it's me, Melissa."

Carmine stopped, glancing back at Melinda and squinting his eyes. He hung up his phone and paused as Melinda and Fredric approached. When they were within 10 feet, Carmine turned and ran, dodging in and out of evening strollers, breaking up couples holding hands. "Radio in for backup! Ask for assistance down on Main Street heading east," yelled Melinda.

By this point, Carmine had crossed the street. The slow pace of the traffic pretty much halted the cars in their path. Carmine jumped on the hood of a shiny white Mercedes convertible. The lady shrieked half in fear and half in anger. Then he hopped onto the hood of a Hummer and slid off the edge, landing on top of a Harley parked on the side of the road. The loud crash of Harley metal scraping on the sidewalk aroused a few onlookers, one of which was the owner. He was a beefy, bald man covered with tattoos. Carmine stopped momentarily to get his bearings and was back up on his feet and moving. As Melinda and Fredric crossed the street, they were joined by the beefy man who looked like someone who was just let out of the joint and looking for revenge on the poor sap who put him there. "He's ours," announced Melinda.

"Not if I get him first," he grinned as he started jockeying for position on the sidewalk.

Their chase led them into the residential area of Manayunk consisting of backyards and small driveways. Carmine was hopping fences and knocking over trashcans in the process, trying to create a distraction. By this point, the worn-out biker, who probably was a heavy smoker, stopped at a corner for a respite from the chase, which was good news for Melinda and Fredric. Carmine was gaining a lead on them and disappeared out of sight a few times. They came upon St. Jehosaphat's. Carmine ran through the church doors, followed by his pursuers. But when Melinda and Fredric reached the inside, Carmine was nowhere to be found. They stopped in their tracks as they looked at the stunning artwork and stained glass on the walls glowing in the lights.

"Where did he go?" asked Fredric.

"Let's split up. You go up to the front and I'll go outside and look around back."

Fredric unsuccessfully searched every niche and corner and then went outside to look for Melinda. She was standing at the base of the steps that came from the back of the church. She was on the phone canceling the backup, because they would not be of much help at this juncture. "I found a gun here. No sign of Carmine, but it might be his. We can take it back to the techs at the precinct and see if they can lift some prints."

Back at the station, Melinda and Fredric met up with Mike Ragney. He was the fingerprint guru of the city. The evidence bag containing the gun was sitting on his desk. "Hey, you lousy bastards, how come you let him go?" joked Mike.

"Well, Mike, you don't have to rub it in," shot back Melinda. "I know you're Mr. Fitness, but we don't all have time to go the gym. Maybe if we sat in our office all day and looked at fingerprints, we could go to the gym, too."

"Okay, let's see what we have here. You have a .45 Glock. I don't want to touch it and ruin any possible prints. I'm going to place some adhesive tape over the barrel to protect it from the fumes. Hassie, did Monsternick tell you about fuming for prints yet?"

"No, but I've heard of it. It's cyanoacrylate, right?"

"You got it, Batman. Just some Superglue. I have this Cyanowand here. All I have to do is load the cartridge and click the starter, which heats up the cyanoacrylate. The esters in the fumed glue will adhere to the oils and fats from the prints."

Mike placed the gun in a three-gallon fish tank that was covered with a thick white film from previous fumings. He closed the lid, fired up the Cyanowand, and inserted it into a little hole at the top. He turned on the exhaust hood and closed the shield over the tank. "You don't want to smell those fumes, man. They'll knock you on your ass like an elephant tranquilizer. Remember when kids used to huff Superglue and paint? Well, same thing. You'll deplete your brain cells and then you won't be able to spell your own name."

They waited patiently as Mike changed the cartridges in the wand several more times. As the fumes went to work, Melinda sat wondering what their next move would be. "Hassie, I think Carmine's going into hibernation. We won't see him at school or his house. We'll have to put out an APB on him. Maybe we can dig up some students at the school and find out where he might be. This gun will have to go out to National Medical Labs for a projectile comparison with the bullet found in Adrian. If we get a match, we'll be one step closer to finding a suspect and making an arrest."

"Alrighty, then," said Mike. "We have some beautiful partials on here. You know there's no such thing as a full fingerprint, right, Hassie? The day you find a full fingerprint at a crime scene is the day I retire. Anyway, it looks like you have some whorls and loops. I can't tell for sure right now which hand they're from or which finger, but I'll send it along to my girlfriend at National Medical. She's my idol. Too bad she's married with three kids, or else..."

"Thanks for getting this done so fast for us, Mike," said Melinda. "We appreciate it. You are a true gentleman, and I use that term loosely. Later."

Melinda and Fredric each went to bed befuddled that evening, for the excitement was starting to creep in as thoughts of how a captured Carmine would solve the mystery about who killed two high school boys and two families could allay their fears of never finding justice for their lost children.

~~ ~~

Vocabulary

Accidental   A subclass of whorl-type patterns that consists of a two-pattern combination that does not conform to any pattern type. It usually has two or more deltas.

AFIS   Automated Fingerprint Identification System. A computer system that electronically compares unknown latent prints to prints in a database of known persons. The computer produces a list of possible matches that meet the satisfying criteria. An official print examiner then determines whether or not there is a definitive match.

Anthropometry   An outdated system in which individuals are identified by measurements of body parts. It was created and used by Alphonse Bertillion in 1883, but proved to be unreliable.

APB   All-Points Bulletin.

Arch   A pattern type in which ridges flow from left to right with a slight rise or hill in the center.

Bifurcation   A forking or dividing of one line into two or more branches.

Central pocket loop   A subclass of the whorl with two deltas and at least one ridge making a complete circuit that may appear in a spiral, oval, or circular pattern. It also has a second ridge recurve either connected to or independent of the first recurve.

Core   The approximate innermost center of a fingerprint impression.

Cyanoacrylate ester   A substance commonly known as "Superglue." It has become a popular method for developing latent prints on surfaces that do not respond well to fingerprint powders or other print-enhancing chemicals.

Delta   A point on a ridge at or in front of and nearest to the center of the divergence type of lines that move away from the center. It may appear as a bifurcation or an abrupt ending ridge, dot, short ridge, or meeting of two ridges.

Dermis   The inner layer of skin that holds the dermal papillae. It serves as the mold for the development of friction ridges upon the epidermis.

Double loop   A subclass of the whorl group with two separate loop formations containing separate sets of shoulders and two deltas.

Enclosure   A ridge characteristic that is made of a single ridge that bifurcates and then reunites to continue as a single ridge.

Epidermis   The outermost layer of the skin that is composed of stratum corneum, and cornified dead cells that fall off like scales.

Friction skin   The ridged skin on the inner surface of the palms, fingers, and feet. It is identified by the absence of hair and the presence of sweat.

Henry classification   The system of classification developed by Sir Edward Richard Henry in 1901. His system is the one accepted by the United States and all other English-speaking countries.

Latent print   Unintentional fingerprints left at a crime scene that are usually invisible to the naked eye.

Loop   A type of fingerprint pattern in which one or more of the ridges enter on either side of the impression; recurve, touch, or pass an imaginary line drawn from the delta to the core; and usually terminate on or toward that same side of the impression from where it started.

Patent print   A print that is visible to the naked eye.

Pattern area   The part of a loop or whorl used in classification in which the cores, deltas, and ridges needed for pattern interpretation are apparent.

Plain arch   A pattern in which the ridges enter on one side of the impression and flow out the other with a rise or wave in the center.

Pressure distortion   The distortion of a fingerprint due to unusual pressure or force when pressing the finger on a surface.

Radial loop   A loop in which the flow of the ridges point in the direction of the thumb.

Ridge characteristics   Minute ridge endings, bifurcation, enclosures, and other details that must match in two prints for positive or negative identification.

Ridge count   The number of ridges intervening between the delta and the core. It is used to subdivide loop patterns.

Ridge trace   A process for subdividing whorl patterns into areas by considering the flow of the ridges from the left to the right delta.

Tented arch   A pattern in which most of the ridges enter upon one side of the impression and flow out the other side with the exception of the ridges at the center, which form an angle.

Ulnar loop   A loop that flows in the direction of the pinky finger.

Whorl   A pattern having at least two deltas with a complete recurve in front of each. A plain whorl has two deltas and at least one ridge making a complete circuit that can be spiral, oval, circular, or any variation of a circle.

Background Information

Use of fingerprints has been in existence for a long time. The Chinese used them as signatures in business deals, and artists signed their names with fingerprints. Before fingerprints were used to identify criminals, a system called anthropometry was used. Anthropometry was a method of measuring and documenting the length of bones, and those measurements, along with the person's information, were kept on file. It was used on several occasions to supposedly identify criminals. As it later turned out, anthropometry was not a valid method of identifying criminals and wrongful arrests were made. Eventually fingerprints were established as demonstrative evidence through the research of several men. An abridged history follows:

circa 1000   An attorney in the Roman courts named Quintilian displayed bloody palm prints that were meant to frame a blind man for the murder of his mother.

1686   Marcello Malpighi, a physiologist noted characteristics of fingerprints, but did not make the connection of how they could be used for individual identification.

1800s   An English naturalist, Thomas Bewick, identified his published books with an engraved fingerprint.

1823   The first paper on the nature of fingerprints and their classification system was published by John Evangelist Purkinji. It was based on nine major types of patterns. The possibility of using them for individual use was not mentioned.

1856   A British officer working for the Indian Civil Service, Sir William Herschel, began using fingerprints to identify criminals. Over the next 20 years, he started using fingerprints as signatures on land titles and jailer's warrants.

1878   Henry Faulds, a Scottish missionary working in Japan, discovered fingerprints on some ancient pottery and began conducting extensive research.

1880   Henry Faulds published a paper about how fingerprints at the scene of the crime could identify an offender. In this particular crime scene, which was a burglary, Faulds actually exonerated a suspect and caught the real one based on fingerprints.

1882   A railroad builder, Gilbert Thompson, put his thumbprint on wage receipts to protect himself from forgeries.

1883   Alphonse Bertillion, an anthropologist identified the first repeat offender based on his invention of anthropometry.

1892   Sir Francis Galton published the first book about fingerprints and how they are used to solve crime. His work led to a fingerprint classification system that would be used in Latin America. Argentina was the first country to replace anthropometry with fingerprints.

1893   Edward Henry, who was the chief of police in Bengal, India, began to add thumbprints to the anthropometric records that he had been collecting. It was not until 1897 when Henry's assistant, Azizul Haque, came up with a comprehensive system for classifying fingerprints. By this point, anthropometry proved to be inferior to that of the fingerprint, so it was phased out.

1901   Anthropometry was replaced by fingerprint identification by the head of Scotland Yard, Sir Edward Richard Henry.

1901   The New York Civil Service Commission made the first systematic use of fingerprints.

1902   The first criminal was convicted on fingerprint evidence in Britain when he left his print at the scene of a burglary.

1903   Fingerprints were used for the first time to identify criminals in New York State Prison.

1918   Edmund Locard, a scientist, suggested that a 12-point identification system be used for identifying fingerprints.

1970   Fuseo Matsumur, a hair and fiber specialist, inadvertently discovered that Superglue could be used to develop latent prints. While working on a case, Matsumur was mounting a hair on a slide and the conditions of the room caused a reaction with the oils and residue in the print with the solution used to make the slide mount. This is now known as "fuming" through the use cyanoacrylate (Superglue).

1977   The FBI introduced the Automated Fingerprint Identification System (AFIS). The early stages of establishing a fingerprint database had begun.

1999   The FBI upgraded their AFIS system to paperless submission, storage, and searching capabilities.

Fingerprints are unique to every individual. Most people assume that fingerprints exist only on the epidermis portions of the skin. Believing this misconception, criminals have tried to erase their fingerprints by scraping the skin off their fingers, burning them with heat or melting them with acid, using a sharp object to cut and scar their hands, or whatever other crazy concoction they could come up with. It does not matter. Fingerprints grow back, and when they do, they are even more unique because of the damage that has been done. Fingerprints actually go down to the dermis level of your skin where it is coded to grow back the same way every time. The best way to get rid of your fingerprint is to cut off your finger and throw it away.

Fingerprints are produced early on in the development of a child. Around six weeks, the hand looks almost like a paddle. This is when skin is beginning to form around the first joint of the fingers, which are called volar pads. Volar refers to the palm of the hand or sole of the foot. Around 13 weeks, the ridges begin to form in the epidermis layer of the skin. This is called friction skin. It is the area of skin that does not grow any hair. That is its identifying characteristic. It also secretes lipids, oils, and fats that help us grip objects. These fluids usually do not really start secreting significantly until approximately 10 to 12 years of age. It is these biological fluids that produce fingerprints on a surface. The nature and life of a fingerprint depends on elements such as temperature, light, relative humidity, and precipitation. Prints that are seen with the naked eye are called patent prints, while prints that cannot be seen with the naked eye, or those that need to be enhanced to be seen, are called latent prints.

Patent prints are ones that are left as an impression. For example, if someone sticks a hand in semi-solid mud, grease, oil, or blood that is almost dry, it will leave an impression that can be photographed and sent to the lab for identification.

"Latent print" is a term that surfaces all the time on TV crime shows. The word "latent" is a Latin word meaning "hidden." To bring forth the print, some kind of physical or chemical processing must be used. This will be discussed in the following section.

There are several kinds of fingerprints. Where your volar pads develop determines what kind of fingerprint you will have. Low pads usually form into some type of arch pattern, high pads usually form into a type of whorl pattern, and pads in the middle usually form a loop-type pattern.

Forensic Implications and Jurisprudence

Once prints are identified at a scene, they must be properly documented in writing, photography, and video. The surface where the print is found determines how it is collected. Some are photographed directly, while others are subjected to development methods. The oldest method for developing prints is through the use of powder. The powder adheres to the sweat outlines of the ridge patterns. In more modern times, a magnetic powder is used to keep prints from being damaged by bad brushing techniques or excessive use of dust. Once the powder is spread over the print, a magnet takes away the excess, leaving a clear, untouched print. Depending on the color of the surface, different color powders and reagents can be used. Caution must be used when using powder because a print can be erased with the stroke of a brush if not done properly. The powder can also get caught between the ridges and distort the print. Not much powder is needed to expose a print. When a brush or duster is used, it should be held between the thumb and index finger and twirled above the print, taking care not to directly brush it.

Fingerprint development had some milestones in the 1970s. Fuseo Matsumur inadvertently discovered that Superglue could be used to develop latent prints. This is now known as "fuming" through the use of cyanoacrylate (Superglue). Superglue packets can be used or a tube of Crazy Glue and a fish tank with a light source for heat and approximately 50 ml of warm water can be used. Today, the Cyanowand has proven to be an effective tool when properly used in the lab.

Hard and nonabsorbent surfaces like glass, car dashboards, guns, and knives are prime candidates for cyanoacrylate fuming. The print then can be enhanced by use of other reagents that fluoresce, providing a better print to photograph.

Iodine fuming is one of the oldest chemical methods for developing prints. Iodine is a solid crystal that transforms into a vapor when heated. It does not pass through a liquid phase, therefore making the transformation one of sublimation. The resulting vapors fill the fuming chamber and combine with the latent print to make it visible. The only downside to using iodine fuming is that the colored prints are not permanent, and need to be photographed immediately. However, prints can be fixed with a 1 percent solution of starch in water and then applied to the print by spraying. This causes the print to turn blue and usually lasts for several weeks to several months.

Another developer used by investigators and technicians is ninhydrin, which is available in aerosol form. It is used on paper surfaces and has been known to develop 15-year-old prints on paper. Ninhydrin reacts with amino acids that are present in perspiration. Over anywhere from 1 to 48 hours, a purple-blue color print appears. It can be quickened if placed on a hot plate at a temperature of 80 to 100 degrees Celsius.

Sometimes locating prints can prove to be the most difficult. Advances in UV light technology have allowed investigators to find and photograph prints without touching them. The Reflected Ultraviolet

Imaging System (RUVIS) locates prints on most nonporous surfaces. When UV light strikes the print, the light is reflected back to the viewer, differentiating the print from the background surface. The transmitted UV light is then converted into visible light, where it can be photographed with a one-to-one lens.

The latest development in fingerprint technology is digital imaging. In this process, a picture is converted into a digital file. The image is composed of many square dots called pixels. Each pixel is assigned a number according to its intensity. These can range from 0 (black) to 255 (white). Once the image is stored, the picture can be manipulated through software that changes the number value of each pixel, which is defined in terms of degree of dimensions. The larger the numbers are for these parameters, the closer the image represents the real thing.

All fingerprints fall within three main categories: arches, loops, and whorls. Approximately 5 percent of the population have arches, 60 percent have loops, and 35 percent have whorls.

Arches can be subdivided into tented arches and plain arches, which are further subdivided into radial and ulnar. A radial pattern points toward the thumb and may display a delta and no recurring ridge, or a delta that is part of a recurring edge. If it is the case that both features appear, there must be no ridge count between the core and delta points. The ulnar arch displays the same characteristics as the radial arch, but points to the pinky finger. In a tented arch, the ridges near the middle thrust upward and converge from both sides of a spine or axis, which gives the appearance of an arch.

Tented arch.

Loops are divided into ulnar and radial loops depending on the slant of the loops and hand on which they appear. The radial is named after radius bone and subsequently points toward the thumb. The ulnar refers to the ulnar bone and points toward the pinky finger.

Whorls are subdivided into plain whorls, central pocket loops, double loops, and accidental whorls. Plain whorls form a circular pattern and have two distinct deltas. The central pocket loop is differentiated by making a line across the two delta points. This line must not touch or cross any ridge formation within the inner area of the pattern. This is the area of the print within the two deltas. Double loops consist of two separate loop formations with two distinct sets of deltas. Only those patterns with two distinct loops fall into this category. Accidental whorls are the few patterns that are so irregular, they cannot be grouped with the other subdivisions. They have

Ulnar loop.

Plain whorl.

two or more deltas and a combination or fusion of two or more types of patterns. This category also includes other patterns that do not fall into any of the previous categories.

Remember, the computer does not match the prints. It takes a professional fingerprint examiner to make the match. The computer usually provides the 10 closest matches and the examiner does the rest. The computer system that is used is called the Automated Fingerprint Identification System (AFIS). It is used across the country and around the world. At crime scenes, all people present are fingerprinted and sent through the system to avoid confusion with prints of the suspect or victim.

Classification and identification are two separate concepts that vary in many ways. The only commonality they share is they both deal with fingerprints. *Classifying* a fingerprint is determined by a mathematical formula based on types of patterns that occur on 10 fingers of an individual and the various subclassifications and divisions within those patterns. It is also established upon the basis of location within the patterns or reference points such as deltas and cores. This classification is usually done once the examiner has a full set of prints to examine. On the other hand, *identifying* the fingerprint compares the individual ridge characteristics such as ridge endings, bifurcations, ridge dots, enclosures, and other distinguishing characteristics. To establish identity, it must be proven that there are a sufficient number of characteristic matches found with frequency through qualitative and quantitative analysis. In the past, this has been the topic of heated debates among judges and lawyers. How many matches are sufficient to call it a match? On an entire fingerprint, there can be up to approximately 150 points to use for a conclusive match. That is why partial prints left at a crime scene can be used to make positive identifications.

Minutiae	Example	Minutiae	Example
ridge ending		bridge	
bifurcation		double bifurcation	
dot		trifurcation	
island (short ridge)		opposed bifurcations	
lake (enclosure)		ridge crossing	
hook (spur)		opposed bifurcation/ridge ending	

Ridge characteristics.

To compare an unknown latent impression with an inked one, the examiner looks for four different elements:

1. The likeness of the general pattern type or similarity in the flow of ridges
2. The qualitative likeness of the friction ridge characteristics
3. The quantitative likeness of the friction ridge characteristics
4. The likeness of the location of the characteristics

After a four-year science degree has been obtained, one can begin an entry-level position in a forensic lab and seek certification as a fingerprint examiner. In order to receive certification, the examiner has to pass a test and meet certain training and experience criteria. Knowledge has to be displayed in the areas of the classification of inked fingerprints, identification of inked fingerprints, and comparison of latent and inked prints. The certification lasts for three years, after which the examiner has to undergo re-certification. Board certification does not guarantee competence, nor does the absence of certification mean the witness is incompetent. It is at the discretion of the trial judge as to who can and cannot testify.

8

The Frustrating Firearm

Two weeks later, Melinda and Fredric once again found themselves standing on the doorsteps of National Medical Labs where they waited to be buzzed in. The detector sniffed their guns and tattled with an annoying buzzing sound. Kristen, the receptionist, was already watching them as they walked over to her desk. "Good morning, Detective Monsternick. Hi, Hassie, how are you?"

Fredric smiled broadly. "I am fantastic today. The sun is shining, the leaves are starting to sprout on the trees, and summer will be here soon. I love it."

"Me, too," she returned, with a sparkle in her eyes.

"We're here to see Dr. Rieders. Could you buzz him for us, please?"

Kristen looked disappointed as she followed Melinda's orders. It was cutting into her flirt time with Fredric. He promised himself that when the case was solved, he would perhaps ask Kristen to have some dinner in a swanky restaurant in the city. She frowned as Dr. Rieders walked through the door.

"Good morning to you," he said in his slightly Hungarian accent. He was the founding father of National Medical Labs and had been there for 40 years now. His passion was his work with firearms. The wall of his office was decorated with several doctorates in various fields of science.

To most in the building, Dr. Rieders was considered the oracle of knowledge. He had testified in many high-profile cases. Integrity was his middle name, for he was a painstaking scientist who took every precaution before sharing his results. It was also customary for Dr. Rieders to share his stories and lifetime experiences. As a sage in the forensic profession, many looked up to him and loved to sit and listen to his stories. However, the newbies and younger scientists were not eager to hear his trials and tribulations of growing up in Hungary and moving to the United States to work in New York City with the medical examiner.

"Well, we won't get much work done here," Dr. Rieders said. "Let's go back to my office so we can speak."

Fredric reluctantly left Kristen in the name of justice. Melinda wanted Fredric to have a comprehensive education in firearms and ballistics, because they were a major constituent of their investigations.

As they sat in Dr. Rieders' office sipping hot coffee, he explained the difference between ballistics and firearm identification. "Ballistics is a term often used in crime scene television shows around the world. For the TV crime scene investigator, ballistics means matching up the bullet with the gun. The real-life crime scene investigator knows that ballistics is the study of the motion of a projectile within the firearm, the motion after it leaves the firearm, and the effect it has on a target. It takes into consideration the size and length of the barrel as well as any obstacles that may lie in the path of trajectory."

"The real art of matching projectiles to a suspected gun is called firearm identification," Dr. Rieders continued. "Along with the projectile comes possible gunshot residue. The firearms examiner is not responsible for GSR, so it is identified by a trained forensic chemist." He held up a .357 snub nose and a .30 odd 6 rifle. "Hassie, if I fired these two guns side by side, which bullet do you think would go farther?"

"I would imagine the rifle's. It's the preferred tool of most big game hunters."

"You are correct," replied Dr. Rieders. "Do you know why?"

"I guess because rifle bullets are longer and have more gunpowder in the casing to send them farther."

"Well, my young lad, not bad. The rifle bullet will go much farther because the barrel is longer. There might be more gunpowder in the rifle casings, but that is not always the case. When I worked as a medic in World War II, these .30 odd 6 bullets whizzed by my head all the time when I had to go out and pick up guys on the stretcher. I can't tell you how many times they missed me out there. One thing that made it easier was the vodka they gave us."

Melinda lifted an eyebrow. "What are you talking about?"

"Detective Monsternick, how do you think we got the stones to walk out on that field with a stretcher? A lot of guys on the front lines went in with a fifth in their jacket pocket. It took away the foreboding thoughts of death and helped them muster some more courage. For me, it made me tipsy on my feet, and they couldn't shoot me because I kept swaying back and forth."

Melinda and Fredric chuckled as Dr. Rieders let out a smile and scratched his gray beard, which was symbolic of his wealth of knowledge and experience. "So, let us return to our lecture about firearms."

"When a rifle is made, the barrel is bored with a drill-like instrument. This is called machining. When the machining is done, another instrument is used to 'rifle' the barrel. What it does is create a spiral in the barrel of the gun. Some of the marks are indentations in the gun, which are grooves. The other marks of significance are the marks that are not indented, which are called lands. Now, since a bullet is usually made of some kind of soft lead, aluminum, or copper, it will easily take on the shape of the barrel. When it is forced through the barrel of the gun, the projectile, as we call it in the science world, takes on the shape of the lands and grooves. This also creates a spin, and the projectile becomes more accurate and is streamlined while traveling through the air. The projectile will travel as far as it can until it either runs out of air speed velocity or is acted upon by some outside force. Now we are entering the realm of ballistics, because we are studying the travel pattern of the projectile and how it is affected once it is acted upon by an outside force, which most of the time is a person or animal."

"So, basically, since a pistol or handgun has a shorter barrel, the projectile doesn't travel as far," observed Fredric.

"Absolutely," answered Dr. Rieders. "This .357 here is accurate to maybe thirty-five yards. After that, the trajectory begins to go down. It usually works in an arc motion. From the barrel to the ground, it follows a descending motion until the velocity slows it down enough to hit the ground. That's why most people hunt with rifles."

Fredric sat pensively for a few moments. "What about guns that are made by the same manufacturer? Won't the lands and grooves be the same in all of their guns?"

"Great question, Hassie!" exclaimed Dr. Rieders excitedly. "Actually, every time the machining and rifling processes take place, small imperfections are created that make each gun as unique as a fingerprint. Another factor that has a direct bearing on making the identification unique is the amount of times the gun has been fired. Yes, some companies have anywhere from two to twenty-six grooves in a barrel, but the more someone sends a projectile through that barrel, the more miniscule changes are made to the inside of the barrel. We used to view the projectiles with a stereomicroscope, but now we can use a comparison microscope, where we line them up side by side. Also, with the use of digital imaging, we can do an overlay or superimpose them on top of each other. It is a spectacular display in the courtroom. Sometimes I feel like Gill Grissom on *CSI*. However, he is way smarter than me and much better looking. He knows every single aspect of forensic science, works in the field, does the job of the detective, and testifies in court about everything! Wow, is he good."

The three of them laughed heartily as Dr. Rieders joked some more about the absurdity of television shows versus real life. It was hard for jury members these days to sit in on a case and not feel the "CSI effect." With all of the glitz and glamour of crime scene-related television shows, jurors were always expecting to see the workings of some fascinating instrument that could positively ID something and use it as definitive evidence against a perp. They also thought that DNA was available from dust on the ground. Sure, DNA is available at crime scenes, but it can be contaminated and rendered useless. Also, some of the collection instruments cost an exorbitant amount of money and cannot be afforded by most police stations or labs. This tainting of jurors sometimes exasperated Dr. Rieders as well as other professionals in the forensic field. The drama was more exciting on television because they did not include all of the testimony and paperwork involved in solving a case.

Dr. Rieders placed some projectile casings on his desk. He asked Fredric if he could pick out the three casings from a 9mm and tell him if they came from the same gun. Fredric fished through the shells as they jingled and clanged together. Eventually, he pulled out three silver shells with 9mm etched in the bottom of the casing. After studying them carefully, he looked at Dr. Rieders and Melinda. "They look pretty much the same."

Dr. Rieders clasped his hands and held them to his mouth. "Do you think they came from the same gun?"

Fredric inspected them some more and scratched his head. "Yes, I think so."

"Hassie, never use the word 'think.' A defense attorney will take you on a wild goose chase and destroy every shred of credibility you have. Trust me, I've been there. What you should say is that you can conclude with a certain degree of scientific validity that the three shell casings match each other. Oh, and by the way, they are from three different guns. Bring them over here and I'll show you."

They walked over to a table with several microscopes. Dr. Rieders placed all three casings upside down so the bottom of the casing was facing up. "First thing you have to look at, Hassie, is the model and make of the bullet. You have picked out three shells that are all produced by the same manufacturer. It indicates that the casing is for a 9mm. Sometimes you may find a specific symbol or trademark, year of manufacture, and possibly the country of origin. The more information you have, the more specific the casing becomes.

"As you look at the bottom of the casing, you'll see marks where the firing pin has struck the casing. Let's look closer." Dr. Rieders placed the casing under the microscope and focused in on the bottom area. "If you look here, you will observe small marks that seem to be striated or impressed across the casing. This is called a breech face marking. The breech face is the area of the firearm that supports the casing during the small explosion of gases. As the casing pushes against the breech face from firing, the marks are indented on the casing. There are three major premises in this job, Hassie, and don't you ever forget what I am about to tell you. It might be the most important thing you ever learn. Are you ready?"

Fredric nodded as he listened intently.

"No two things are made the same way, break the same way, or wear the same way. The gunshot residue changes from batch to batch. It only shows up on the molecular level when you run a sample through the ICPMS. Same thing with a screwdriver. The more I use it to pry things open or push a screw through a piece of wood, the more I change its outward appearance. You will never find the exact same marks on two similar screwdrivers. Another example, take twenty new pencils. Break every single one of them, and you will be able to see with the naked eye that they do not break the same. These are three major concepts that rule my job. You should always have them in the back of your mind when you are investigating a case. Question everything, young man. It will only help you create more questions that will lead you to the pursuit of justice through science."

Fredric nodded in agreement. "That makes a lot of sense. I never thought of it that way. No wonder you are called the 'oracle of knowledge' around here. So the breech face markings are another possible way to identify and match a firearm to a projectile?"

"Yes, but that's not all. Let's think about the firing pin. Take a look at the area where the pin has struck, initiating the spark that causes the projectile to fire. Then look over here at the gun I have taken apart. I placed the actual firing pin under the microscope."

Fredric examined one, then the other. After he finished, Dr. Rieders used his digital imaging to superimpose the two on top of one another. It was a perfect match, or at least to a high degree of scientific certainty. Dr. Rieders smiled as he saw the light bulb go on over Fredric's head. "So you see, the firing pin makes an impression on the casing. It has several class characteristics such as a hemisphere, elliptical, rectangular, wedge, semi-circle, or even a U shape. It all depends on the manufacturer."

"What are class characteristics?"

"Class characteristics are ones that are common to a particular family or group of items, such as guns manufactured by one company. When we look at the firearm in depth, we are talking individual characteristics that make it individual or unique to any other firearm. These may also be referred to as accidental characteristics. I like to refer to them as the owner's signature on their firearm."

"Is there anything else that helps make a match?" inquired Fredric.

"Yes, there are several things. One is the ejector marks on the casing. This is when the casing is ejected from the weapon after it is fired. Those marks are unique as well. The other is the gunshot residue and its chemical composition. All casings or shells are packed with some kind of barium, antimony, or lead. Then a primer is added. When the firing pin strikes the casing, a spark is created, which ignites the primer, and the projectile is fired out of the barrel. Gunshot residue shows up in many places. First, it's in the casing after the projectile is fired. If someone is standing close enough to the gun as it's fired, the GSR will appear on the person's clothes or skin. Anything beyond ten feet is difficult. Also, the shooter will have some residue on his or her hands because of the exploding gases. It shows up really well under a UV blacklight. It has to be a blacklight, though, because regular UV light doesn't have the same wavelength.

"The best way to collect this material in the field is in the articles of clothing or fabric that might be exposed. If someone is shooting from a car, it might show up on the upholstery. Depending on what kind of collection kits you have available, you might have to use a swab with distilled water, or even better, GSR tabs. These work great on skin, because it's like putting a stamp on someone, but instead of a stamp of a happy smiley face, these tabs are sticky and collect the residue particles. Then we can run them through a GCMS or the ICPMS. The latter is proving more effective. From there, sometimes we can match the chemical composition of the GSR to a box of bullets found in a perp's house with the same lot number. Of course, the naysayers will not buy into it, so we are conducting more extensive research to prove them wrong. One of our scientists is publishing a paper in the *Journal of Forensic Science* next month. I am sure you can't wait to read it."

"No, I can't," replied Fredric.

"So, Dr. Rieders," interjected Melinda, "Do you think we can run a test on the gun we found and the bullet we retrieved from Adrian?"

"Sure, we need to go over to the firing room so we can acquire a few samples."

Melinda and Fredric walked next door, which was the end of the building. As they entered the room, they saw a 55-gallon drum full of water sitting in the middle of the room. Dr. Rieders opened a drawer and pulled out a box of bullets that had the number .45 written in boldface on the side of it. He loaded the suspect .45 Glock that Melinda found outside of the church after their escapade with Carmine. "You had the prints lifted from there, right?" asked Melinda hesitantly.

"No," said Dr. Rieders, "I was going to put my prints all over the gun before I sent it over for an AFIS run. I'm eighty-five, but I still haven't lost the brain cells that tell me to have evidence examined before I touch it. My Alzheimer's seems to come and go, especially when my wife asks me to do something trivial."

Melinda burst out laughing. What a silly notion to ask one of the world's top forensic scientists if he had handled a piece of evidence properly. Fredric knocked on her head and asked if anybody was home. Then he pretended to look in her ear and searched for his hand on the other side indicating nothing was there. "All right, that was a stupid thing to ask," she said.

Dr. Rieders put on some earphones as well as a pair of goggles and Melinda and Fredric followed suit. He fired three test shots into the 55-gallon drum. He then explained to Fredric that they needed the drum to stop the bullets and allow them to sink to the bottom. The next crucial step was the retrieval of the bullets. Dr. Rieders pulled out a long stick with what appeared to be gum at the end. "You never want to use anything but some sticky putty to collect the expended projectiles. If you use something hard, it might damage the projectile and add marks that are not consistent with the ones in the barrel, hence compromising your evidence."

He retrieved the three bullets and brought them back to the lab. He placed the first one under the comparison microscope with the suspect projectile from Adrian. He turned the knobs and pulled up the two images on his computer to superimpose them. It was not a match. The suspect projectile had six grooves, while the test projectile had eight. Just to be sure, Dr. Rieders compared the other two samples, but to no avail. "Well, there you have it, clearly not a match. I would say with a certain degree of scientific validity that this is not a match. Sorry, guys, you'll have to find another gun for me. I'll place the evidence back in the log room and seal it up for you."

Melinda and Fredric both let out a big sigh as they nodded in discontent. The best possible link to Carmine DiGuiseppe and the killings was now a thing of the past. The only evidence they had linking him to anything suspicious was the mention of his name in the Drater Club notebook. The only thing they might be able to nail him for is possession of a firearm without a permit or registration, and they did not even know it that was plausible, for the prints had not yet been run through AFIS. They walked out feeling dejected, until Fredric saw Kristen as they were leaving, which did not do anything for Melinda.

Vocabulary

Ballistics   The study of the travel pattern of a projectile and how it strikes a target.

Blowback effect   The result of close or direct contact of the barrel on the skin. Tissue and blood particles will appear in and around the barrel of the gun.

Breech face markings   Markings that are made when the cartridge hits the back of the gun as a result of the exploding gases.

Comparative micrography   Use of microscopic instruments for comparison.

Ejector markings   Markings that are made when the cartridge is ejected from the gun.

Firing pin   A pin behind the barrel of a firearm that strikes the container of explosive (primer) to make the cartridge fire.

Grooves   The unindented parts of a grooved surface; for example, a ridge between grooves in the bore of a rifle.

GSR   Gunshot residue. Any residue that is left on a surface, clothing, or skin after a gun is fired.

Lands   The indented parts of a grooved surface; for example, an indentation between ridges in the bore of a rifle.

Machining   The drilling process of creating a gun barrel.

Rifling   The cutting of spiral grooves in the barrel of a gun.

Stippling   A pattern of gunshot residue that results from close-range gunshots. The GSR appears on the surface as little dots or a stippling effect.

Trajectory   The path of a high-speed object through space.

Wadding   Material used to hold powder or shot in a gun or cartridge.

Background Information

There are several types of firearms. They can be placed into the following categories:

- Revolvers
- Pistols
- Derringers
- Rifles
- Shotguns
- Machine guns
- Combination of rifle and shotgun

A common misconception is that ballistics and firearms identification are the same. They are actually two separate categories. Ballistics is broken down into three main parts. First is the interior, which is the study of the motion of the projectile within the barrel of the firearm. Next is the exterior, which is the study of the motion of the projectile after it leaves the barrel of the firearm. Last is the terminal, which is the study of the effect of the projectile on the target. Firearm identification deals with the identification of ammunition and components of that ammunition as having been fired by a particular firearm that is unique to all others.

Every bullet fired from any gun has its own identifiable characteristics. Rifle and pistol barrels are produced by a process called machining. After they are machined, they are rifled. In this process, internal grooves are bored into the barrel of the firearm, creating a twisting, almost spherical, shape. This determines the caliber of the firearm, the number of lands and grooves, the direction and angle of twist, and the depth of rifling. This gives the projectile a spin and creates more accuracy when firing. For the scientist, it makes each projectile as unique as a fingerprint. Some companies make firearms that are similar, but all are unique in some shape and form. Also, as the weapon is fired more, the barrel shows slight derivations in the marks that are impressed on the projectile.

Gunshot residue (GSR) can be found in several places. It occurs when a firearm is fired at close range. Beyond a range of approximately 10 yards, not much GSR appears around the entrance of the projectile, whether it is on a surface, directly on the skin of the victim, or on clothes. As the distance of the shooter increases, the trajectory of the gases and projectile decreases.

In cases of homicides and suicides, there is usually a "stippling" effect on the skin or clothing, which looks like little dots around the area of the entrance of a projectile. This usually indicates that the barrel of the gun was not in direct contact with the skin. Stippling can also be

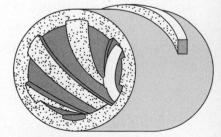

Interior view of gun barrel.

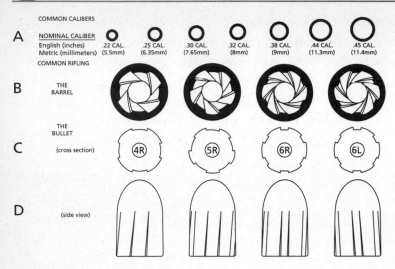

Class characteristics of a barrel.

seen on clothing or surface areas where a projectile has entered. Another artifact of GSR is from a gunshot in which the barrel of the gun is in direct contact with the skin. This happens due to the "pocket" effect. The explosion of the gun causes the skin to open and allows the gases to form a pocket between the skin and existing muscle and tissue. The gases and powders consume this space and separate it, creating a pocket. The force and speed cause the skin to rip or tear and cause a star shape on the skin. In instances like this, GSR can be found inside the wound and around the damaged tissue.

Another area to examine is in and around the barrel of the gun. After the shot is fired on a contact surface, a blowback effect is created. A vacuumlike process occurs, and the barrel of the gun literally becomes a vacuum that sucks in skin, tissue, muscle, and possibly shards of bone. It is easily seen under microscopic examination, where it can be documented and collected.

The person shooting the firearm comes in contact with gases from the explosion. Until the hand is washed, GSR adhere to the skin. Sometimes even after thorough washings, traces of residue remain behind. Under alternate light sources, GSR has a glowing effect. The suspect area can be tested for GSR by using a kit designed for officers to use. These are field tests and can be used right on the spot. Swabs or gunshot residue tabs can also be collected and sent to the lab for examination.

Forensic Information and Jurisprudence

Some of the attributes that make firearms unique are the barrel, chamber, frame, slide, bolt, extractor, ejector, magazine, and feeding ramp. At the lab, the questionable gun has to be test-fired. If the suspect bullet can be identified to the company that made it, then those kinds of bullets should be used in the test. The gun is loaded and then fired into a steel tank of water. It slows the speed of the projectile and allows it to safely rest on the bottom the tank. Then it is retrieved either by hand or by a pole with putty, which adheres to the projectile. This prevents any damage to the bullet. It is then placed under a comparative microscope. Both the suspect bullet and the test bullet can be compared. The scientist can then decide whether it is a match.

The impressions on a projectile are created when two objects forcibly contact each other. In the case of a firearm, it creates lands and grooves. Lands are areas that are indented on the bullet. Grooves are the areas that stick out. This sounds backward, but it is the gun barrel that causes these marks. Therefore the grooves and lands in the barrel have the opposite effect on the bullet. There may be any given number of lands and grooves on a bullet.

In the case of shotguns, most are not rifled, so they cannot be identified. They also shoot pellets and small BBs that are difficult to identify. Slug guns, shotguns with rifled barrels primarily used for hunting, can be easily

Lands and grooves on a bullet.

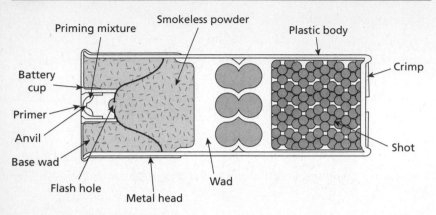

Cross-section of a shotgun shell.

identified as a rifle or handgun. Most "sawed-off shotguns" do not have rifled barrels and cannot be identified. The point of sawing off a barrel is to give the gun a wider spread, meaning the pellets disperse in a wider pattern at closer range.

When identifying a shotgun, the wadding is used for comparison. The wadding is used to keep the pellets together in the shell. It comes flying out when the gun is fired. Shell casings also have distinct markings. Firing pin indentations are produced when the hammer strikes the primer and starts the reaction that fires the bullet. Some new guns, such as automatics, auto-loading shotguns, and semiautomatic rifles, are hammerless, but still possess a firing pin. Breech face markings are caused by the burning gases inside the bullet that force the cartridge against the back of the gun.

Firing pin indentation.

In automatic firearms, extractor markings and ejector markings are found. Extractor markings are made when the next bullet is loaded into the firing chamber. Ejector marks are made by the mechanism that ejects the spent bullet from the gun.

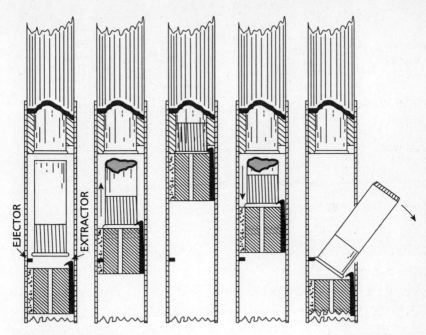

Extraction of a projectile.

All of these marks can be seen using a process called comparative micrography. It requires the use of a comparison microscope. This way the scientist can see both specimens and match them up. Digital imaging is now paving the way for labs across the country.

If a person is interested in becoming a firearms identification or ballistics expert, he or she needs a bachelor's degree in one of the sciences, preferably chemistry, or a degree in forensic science or criminalistics. On-the-job training in the crime lab takes place for a few years until he or she is ready to work on his or her own.

9

Monk's Money

There was only one more week until school let out for the summer. The Friday-morning chaos was business as usual for Melinda and Fredric, who sat in traffic on the Schukyll Expressway. Running out of leads, the two sat quietly, reflecting on their thoughts. Fredric broke the silence first. "Monsternick, let's take a look at the Drater Club notebook. Maybe we can find some more names in there or something."

"My thoughts exactly," replied Melinda. "Before we get to that, we need to speak to the parents. They've had a few days to deal with the deaths of their sons and the funerals are over. I think we should take a ride over to Adrian's house first, then we can head over to Phillipe Mindrago's house."

Adrian Belkoy had lived in one of the row homes on Green Lane. The lackluster landscaping in the front yard was indicative of the sullen mood of the household. The perennials and shrubs were overgrown, and the grass was high enough to provide coverage for small children. A broken lawn chair sat next to a tarnished birdbath now moldy from stagnant water. The yard was guarded by a lawn gnome that was missing its left arm and a scraggly mutt who had not seen a splash of water in weeks. Melinda and Fredric approached the house as the dog began to bark loudly and the fur on her back stood up. Fredric put out his hand and immediately gained the affection of the lonely dog, who started licking his hand. When they knocked on the door, a middle-aged woman answered the door in raggedy sweats with hair to match. "May I help you?"

"Yes," replied Melinda. "My name is Melinda Monsternick, and this is Fredric Hassloch. We're from the Philadelphia Police Department. We're investigating the case of your son. Would we be able to have some of your time?"

"Sure, come in. I'm making some coffee. Would you like some?"

"No, thanks," answered Melinda, while Fredric gladly accepted a cup with cream.

"We're terribly sorry about what happened, and please accept our most sincere condolences. I know this is hard for you right now, but what can you tell us about your son?"

Mrs. Belkoy was visibly fighting back the tears as she took a deep breath and exhaled hard. "I still can't believe he's gone. My only son was on his way out of this dump and God took him away. Plus my mother passed away about a month ago. How do you go on living? I feel as lifeless as scum on the bottom of the pond. I have nothing left to live for."

"We understand, Mrs. Belkoy," started Melinda.

"No, you don't," she snapped back. "Everyone loved him. He was popular, he was an amazing athlete, schools were sending him gifts left and right, he was on his way to the top."

Fredric stopped her. "What do you mean by 'gifts'? Were they like cash gifts or just so-called presents?"

"He never received any actual money, but we would get boxes from these universities filled with all kinds of clothes and footwear. One school gave him a five-thousand-dollar gift certificate for the Gallery in Center City. He had free tickets to sporting events, and one school even offered to set him up with stock options. I don't know if that counts for cash, but it goes on and on."

"That stuff can be cashed in," stated Melinda. "How did Adrian handle all of this?"

"He gobbled it up. He thought is was the greatest thing in the world and told me how he would be rich one day and come back and buy me a new house out on the main line and take care of me and his father."

"How was his relationship with his father?" inquired Fredric.

"They were best buddies. They loved sports and went to all kinds of games. If I didn't know any better, I would say they were best friends."

"Did Adrian ever mention anything to you about a Drater Club?" asked Melinda.

"No. What is that?" said Mrs. Belkoy, puzzled.

"It's a notebook we found in his locker at school," said Melinda. "It was a running log of entries for different sporting events that we suspect Adrian was betting on. Maybe the book was not his, but it was in his locker. We have it at the office if you'd like to come in and see it."

"That doesn't sound like him. He loved sports, but where would he get the money to do that? If you haven't noticed, we aren't that well-to-do."

"He probably cashed in some of his gift certificates," Fredric replied. "Sometimes they're set up that even if you make the smallest purchase, you receive cash in return."

"Well, how much money are we talking here?" asked Mrs. Belkoy.

"Possibly tens of thousands," said Melinda.

Mrs. Belkoy sat for a few seconds with a look of bewilderment. She covered her face with her hands and burst into sobs. "Do you think he was killed by someone for not paying his debts?"

"It's possible, Mrs. Belkoy," offered Melinda. "That's what we're trying to figure out. Any information you can give us would be extremely helpful. We want to do everything we can to help."

"I can't talk about this anymore," cried Mrs. Belkoy. "You need to leave for now. Please come back tomorrow or Sunday when my husband is home. I can't do this right now. I'm sorry."

"It's all right," consoled Melinda. "We completely understand. It's a true tragedy. If you can think of anything else, please give us a call. Take care and let us know if there is anything we can do for you."

"You can find the bastard who did this and send him to hell," Mrs. Belkoy yelled as she broke into a hysterical fit of tears.

"We are truly sorry," said Melinda. "We'll be in touch."

As they strapped on their seat belts, there was a comfortable silence and Melinda and Fredric read each other's minds. Fredric shook his head as he let out a deep breath. "Wow, that was a first for me. Please tell me it gets easier after the first one, Monsternick."

"Well, Hassie, I can't say it ever gets easier, but you learn to create your own defense mechanisms that give you the strength and courage to face these situations. This is by far the crappiest part of the job. It makes you appreciate everything you have and the loved ones who surround you. Are you ready for round two at Phillipe Mindrago's house?"

"Yes," Fredric nodded. "Just take the long way there and let me gather my emotions."

When they arrived at the Mindrago residence, they quickly picked up on the outward appearance of money. The house was surrounded by wrought-iron gates and a beautifully paved brick driveway. Parked in front of the house were a red convertible Mercedes and a white Cadillac Escalade, both polished and shining in the midday sunlight. What they did not realize was that the house was situated on an acre of land, which was unheard of in the neighborhood of Roxborough. Mrs. Mindrago greeted them at the door dressed in a Polo blouse and white Capri pants. "Yes?" she said with an ostentatious squint of the face.

"Hello, Mrs. Mindrago. My name is Melinda Monsternick, and this is Fredric Hassloch. We're from the Philadelphia Police Department. Would we be able to ask you some questions about your son Phillipe?"

"What do you need to know, other than he was the next Lebron James, and now we'll never see his fame and fortune come to fruition? It's highly unlikely that you've done anything worthwhile in this investigation, or you would have a suspect by now."

Taken somewhat aback, Melinda tightened her lips and cleared her throat. "Mrs. Mindrago, I can surely understand your state of frustration, but please, we're here to help. If you don't cooperate, it just makes it harder for us to do our job. Please accept our most sincere condolences. We want to do everything possible to solve this case and bring closure to you and your family."

"Well, you haven't done much yet," she retorted.

"What's going on here?" questioned a voice from the stairs. It was Mr. Mindrago, who was decked out in his Armani suit with a black t-shirt underneath.

"We're from the police department, sir. We'd like to ask a few questions about your son, if that's okay with you."

"We're very upset about what happened to Phillipe. If you don't solve this case, we're going to sue the city and have your jobs."

Melinda, not one for pretentious snobbery, decided her nice gal-approach was not going to work here. Their hostility was blatant, and the only way to deal with that was to return the favor. Melinda had dealt with many people and learned how to read body language and pick up on cues that would reveal their true selves. She poised herself and prepared to lay her cards on the table. "Well, sir," started Melinda. "I wish that were true for your sake, but it looks like you have enough money already. May I ask what you do for a living?"

"I don't have to tell you anything. We inherited our money from my grandmother in Europe. I work in real estate and I do very well."

Melinda realized that even though Mr. Mindrago wanted to act tough, he also wanted to show his cocky, arrogant demeanor that would only expose him for the jerk he was.

"I'm sure you are very successful, sir," replied Melinda. "Do you know anything about a Drater Club that Phillipe was involved in?"

"A what?" replied Mr. Mindrago. He looked at his wife, who was just as confused as he was.

"We never heard of it. Why are you wasting our time? Why don't you come back when you have something substantial to say. Otherwise, we are done here." With that, Mr. Mindrago moved toward Melinda and Fredric, forcing them onto the doorstep where they were met in the face with the door.

"I guess that went well," said Melinda. "Talk about your arsholes. Fredric, we'll have to keep our eyes out on those two. We should probably run a check on them and look into their background. It might turn up something."

ᕞ ᕛ

Back at the station, they pulled out the Drater Club notebook from the evidence room. They both slapped on a pair of gloves because it was supposed to go to the lab for a possible fingerprint check. "Are we going to send this out for prints or what?" questioned Fredric.

"I don't know. After we look through, I'll call Chief and ask her."

As they opened the book, they started looking for any kind of patterns or similarities in the bets, when they were made, who was picked, wins, losses, and of course, other potential names. The bets started in the fall with some college football and professional football. Next to each team, the spread was listed with other information about some of the players and who was injured for the week. It seemed that Adrian was doing well with his bets and he kept a running tally of his winnings. By the time the Superbowl came around, he was up to $10,000, which he displayed with dollar signs for each thousand. For some odd reason, he only bet $2,000 on the Superbowl and still won. Fredric noticed something peculiar about each week. "Monsternick, I'm seeing these entries titled, 'Jo Jo's.' I've never heard of a team by that name, so I'm assuming it's something else. There's nothing listed except for a win or loss."

As they flipped through more pages, they found one entry that said "Royal Flush" with exclamation points. "It must be poker," exclaimed Fredric. "It's probably the hardest hand to get, but it does happen once in a while. So he was playing poker pretty much every Friday since before his death."

"Hey, look, here's one with Phillipe Mindrago's name in it!" Fredric exclaimed. "It seems he lost a thousand that night. Monsternick, we have to find out who this Jo Jo character is. Maybe the boys owed him some money and he killed them. I know it's a long shot, but hear me out. What if Jo Jo is connected with Carmine? They all play cards together and Carmine and Jo Jo are the masterminds. They play cards for fun, but run other scams and illegal operations to make their dough. Carmine did have a piece on him, and you said it yourself, he's probably not afraid of jail time. Maybe Carmine does the setup and Jo Jo finishes them off. He could be our killer!"

"Hold on, Hassie. You're throwing this at me too fast. So, let's see, Carmine does the talking and Jo Jo does the killing? You might have a good hunch there. We need to find out about Jo Jo and see if we can set up a surprise visit on Friday night. Hassie, if this works out, I will be highly impressed. You might even make the front page if you crack this one. Then you get your Batman pin."

"My what?" asked Fredric.

"You know, your little pin with the Batman insignia. It's the chief's way of saying you did a great job. You don't have to wear it, but it's recognition."

"Great, Monsternick. So let's cruise over to the school. We'll check out the rest of this when we get back. We have about one more hour before they get out. Some of the kids will have after-school sports, so maybe we can dig something up on our new man."

The afternoon sun was hot and the humidity made it feel 15 degrees hotter. As they got out of the car, Fredric's shirt clung to his skin like static. He let out a stifled sigh as he noticed a brand-new 2006 red Corvette in the parking lot. "Look at that. I want one of those someday. Those things are so rad."

"Relax, speed racer. Maybe if you never get married and don't have kids, you could afford it, otherwise you're looking at a Matchbox for Father's Day. Wait, isn't that Principal Monk's parking spot?"

"Yeah, it is," replied Fredric. "It has his name above it, so it must be his. Nice ride."

In the office, Principal Monk was waiting with his door open. "Greetings, detectives, how may I help you?"

"Say," said Fredric, "that's quite a ride you have out there. What was your impetus for buying that?"

Principal Monk's eyes kind of shifted away. "I, uh, wanted to do something for myself for once. I had some extra cash I just came into and decided to spend it."

"What, did you win the lottery or something?" asked Fredric.

"No, I, uh, cashed in some of my stocks. So, what can I do you for today? Carmine is not here."

"We expected that," replied Melinda. "We were wondering if we could speak to some of the students and teachers."

"That shouldn't be a problem. My secretary will help you with anything you need. I have to run, I have a meeting at Central Office. Let me know if I can help in any other way."

"Thanks, Principal Monk," gestured Melinda with a wave. "Hey, one more thing before you go."

Principal Monk stopped dead in his tracks and turned with a look on his face like he'd just met Dr. Death. "Yes, what is it?"

"Do you know any students by the name of Jo Jo?"

"No, not that I know of, anyway. It could be a nickname or something. I gotta go. Take care."

With that, he walked out of the office and cruised out of the parking lot in his new toy. Melinda focused her attention on the secretary and asked if she could have the names of Adrian's and Philippe's teachers. They would start there and hopefully find some students.

After the secretary wrote down a few names of teachers who had their last period free, she grinned and walked away. Fredric and Melinda walked into the hallway and their eyes met. "What the hell was that about?" asked Melinda.

"I don't know, but there was something wrong with Monk. I have never seen him that jumpy. I want to say he was almost suspicious. Did you pick up on that, too?"

"I sure did," she said. "Do you think he might have anything to do with this?"

"It's a possibility, but it doesn't add up. Then again, I don't know anything about the guy. He could be the Green River Killer for all I know. We'll run a check back at the station along with the Mindragos. For now let's focus our sights on his teachers and hooking up with some students. It says here that his math teacher is down here. Let's see what we can get out of him."

When they walked into the room, Mr. Simons looked up from his desk. "Can I help you?" he said shyly.

"My name is Melinda Monsternick, and this is my partner, Fredric Hassloch. We're from the Philadelphia Police Department. We were wondering if we could ask you some questions about Phillipe Mindrago and Adrian Belkoy."

"Why, yes," Mr. Simons said, as he visibly started to shrink in his chair. "They were both incredible athletes, but Adrian was the mathematician. His specialty was dealing with scenarios in the area of probability. Of course, all of his scenarios would have to do with sports, but seeing the guy was an athlete, it makes sense. Do I need to come down to the station or anything? I really don't like confrontation."

"Sir, you have nothing to worry about," said Melinda. "We're only here to ask you a few questions. Of course, if there is something you are withholding, then we would have to take you to the station. Is there anything you want to tell us?"

"No, not really. Like I said, I don't like confrontation and talking about this is making me nervous. Please don't take it the wrong way, but you are making me really nervous."

"Well, how do you stand up here every day and teach?" asked Fredric.

"I teach seniors, so most of the time they're working on their senior projects. I don't teach much. I'm just here to provide guidance for students."

"All right," said Melinda. "Have you ever heard of a Drater Club?"

"No, I haven't. I don't sponsor any after-school activities. Sorry."

"Thank you," said Melinda. "We won't take any more of your time. You have a great day, sir."

They walked out the door, and both agreed that although this guy was weird, he did not seem the type to murder his students and entangle himself in the world of gambling. If he thought the Drater Club was an after-school activity, he definitely had no clue about the extra curriculum in high school.

By this point the regular school day was over and the halls were silent as a few teachers could be spotted leaving their classrooms and locking up. Melinda and Fredric headed outside, where students filled the fields with lacrosse practice, track and field, softball, and baseball. Not sure where to start, they headed to the softball field first because it was the closest. The girls were busy at batting practice. They watched as the pitcher whizzed the ball in with some heat. "That girl's pretty good," Melinda commented. "My best friend was a pitcher when she was in high school. She broke all kinds of records, one of them being pitching an underhand fastball at eighty miles per hour."

Fredric nodded with a grin. "I'd hate to be the catcher on that team."

"Well, actually, this is a pretty funny story. One time she was over at my house. My brothers were busting on her and claiming how softball wasn't a real sport. She decided to invite them outside for a game of pitch and catch. So my older brother Karl decided he would be the catcher, and my younger brother Chris would be the batter. Well, her first pitch clocked Chris in the elbow. He held back the tears and egged her on to throw another one. She threw her second pitch even harder. It went down and low, bounced off the ground, went under my brother's mitt, and hit him right in the balls. Needless to say, he was writhing on the ground in pain and talking like Mickey Mouse. Gosh, I remember that one like it was yesterday. After that, they never said anything to her again, except when Karl asked her to his senior prom."

"Where do you come up with these whacky stories, Monsternick? You have one for every occasion, don't you?"

"When you grow up in the boonies of the Poconos, you never know what's liable to happen. Let's say cow-tipping was more fun than the bowling alley on Friday nights. So, let's get back to our reason for being here. There's a lady over there who looks like the coach. I'll ask her for some help."

The softball coach was a beautiful young lady with long blonde hair tied up in a ponytail. She looked more like Barbie than the coach. The only thing separating her from the students was the whistle resting around her neck. Melinda approached her and offered a handshake. She introduced herself and Fredric.

"Nice to meet you," the coach said in a squeaky voice. "It is so horrible. Let me get one of the girls to help you. Aja, can you come over here and help us out?"

Aja was more than willing to give Melinda and Fredric all of the gossip about Adrian and Phillipe and their social life. She must have thought she knew everything, but obviously not, because she never mentioned sports, betting, or the boys' athleticism. However, she did know their friends and who was dating who, when they broke up, and who was cheating on who. Twenty minutes later, the only valuable piece of information they received from Aja was that one of their good friends was on the lacrosse team. His name was Will Doering and he wore the number 33 during practice.

The boys were in the middle of warm-up exercises. As Will was crossing the field practicing his hi-steps, Melinda and Fredric caught his attention and waved him over. He looked behind him and then pointed to himself. Melinda nodded her head and he came trotting over. The sweat was already dripping off his chin and soaking his jersey.

"Hi, are you Will Doering?" questioned Melinda.

"Yes, I am. What's up?"

"My name is Melinda Monsternick and this is Fredric Hassloch, and we're from the Philadelphia Police Department. We were wondering if we could ask you a few questions about Adrian Belkoy and Phillipe Mindrago."

The color drained from Will's face as he stared at the ground. "So, like, what do you need to know?"

"We're investigating the case and trying to find any leads," said Melinda. "You know, like, is there anything outside of school other than athletics they might have been involved in? Did they know anyone who would want to possibly harm them? Were they involved with anyone who did something like, maybe, taking on bets or perhaps some kind of betting? Would you know anything about that?"

Will instantly started sobbing. "I can't take this anymore! I lost two of my good buddies. I told them not to get involved with that crap. It's bad news. My uncle was into that stuff and he buried himself so far in debt, he had both his legs broken. I hate when people do stupid stuff they can't get out of. If I tell you stuff, can you not tell anyone I said it?"

"Maybe," replied Melinda. "It depends if we need to eventually call you into court to testify. Chances are, we won't. We would appreciate anything you could tell us, and we won't have to mention it to anyone right now."

By this point some of the players noticed Will standing on the sideline speaking to Melinda and Fredric. Wanting to avoid any further possible scandal, he spoke quickly. "Look, they would run with this guy named Carmine DiGuiseppe. He goes to school here, but he rarely shows up. Anyway, he knows a lot of people who are into the betting thing. Adrian and Phillipe didn't like to be seen with him around school, but they hung out at night. Every Friday, they would drive up to the northeast section of Philly and play cards with this guy named Jo Jo. I don't know what he looks like, but I do know that the address is 420 Roosevelt Boulevard. I only remember the address because one of them made a joke about his address being the same as the 420 that kids use as a code word for smoking pot. I don't do that stuff, and I don't think they did either, but they did a lot things I didn't know about and don't want to. I gotta go before people start to get suspicious. Please don't contact me or come back to school. I just want this stuff to stop."

"Thanks, buddy," replied Fredric with a pat on his back. "When you go back over to your teammates, you tell them we are scouts from Virginia and we wanted to speak to you about setting up an appointment to visit our school. Okay? Best of luck, son, and you have done you and your friends a great justice by sharing that information. Bye."

It was now four in the afternoon. The two had enough time to go home, shower, grab a bite, and head up to northeast Philly for a date with their new friend, Jo Jo.

10

Jo Jo's Joint

Now that late spring had made its appearance, the sun flashed the sky with a pinkish hue. Melinda and Fredric found themselves sitting in the parking lot of a strip mall that was across the street from 420 Roosevelt Boulevard. The car windows were rolled down and a light breeze wafted through the car every few seconds. Both detectives had a fresh refill of mocha lattes from the Dunkin' Donuts behind them. It might be a while before any action showed up, so the two of them reclined their seats and assumed the lazy position for the time being.

"Let's make a bet, Monsternick," offered Fredric. "I'll put twenty bucks down that nothing happens until ten thirty-two."

"Why such a precise time?" she asked. "Why not round it off to the nearest hour or half-hour?"

"I don't know. It makes it more interesting. Are you in?"

"Sure, I'll choose nine forty-seven. I think if it's poker night, these knuckleheads will want to start as soon as possible to try and win their share of cash. That's assuming they have a poker face. So, let's see, right now it's nine fifteen. We have a half an hour to relax and get ready. There's a Payless Shoes behind us. Do you want to go over and buy some running shoes? You'd be surprised how fast some of these guys are. When I first started out, we had a bust one night and I came through the back door. This dude comes flying down the steps and as he turns the corner, he runs me over and takes off through the back door. Well, I jumped up and bolted off. Another one of the guys on the SWAT team joined in. I was

147

in pretty good shape, and the SWAT guy was at the top of his game. We were left in the dust. This guy was hurtling fences and never turned back. He was ripped and built like a brick house. And when you get one on meth, forget it. These guys would beat anyone in the forty-yard dash, hands down."

"I think I'm pretty fast," said Fredric. "I played football in high school and I was the second-fastest on the team. I'm always up for the challenge. Plus, I don't mind tackling a punk or two and scraping them up."

"Well, you might get your chance." Melinda motioned a nod across the street at a young man entering the apartment. He had a shaved head with a thick goatee and wearing a Metallica shirt with the sleeves cut off. He had tattoos up and down his arms, and his wallet was secured in the back pocket with a chain connected to his belt. He was the stereotypical biker-type you would expect to see at one of the seedy bars in the lower end of the city. Neither Melinda nor Fredric recognized him.

"Do you think that's Jo Jo?" asked Fredric.

"It very well may be. The house was dark up to this point. I see a few lights coming on. Could be his roommate or maybe a friend. Let's chill for a bit. The time now is nine twenty-five, so if the rest of the gang arrives within the next forty-five minutes, I win. We said forty bucks, right?"

"No, Monsternick, just twenty. I'm not that rich."

They waited for another hour, but saw no one coming to or leaving the house. Then a Lincoln Navigator showed up and six guys exited the car. They looked like kids because of their baggy jeans, long t-shirts, and baseball caps pulled down low over their eyes. The last person to get out of the car was a well-built fellow wearing light khakis and a pink-collared shirt. He turned to survey the surrounding area, and Melinda and Fredric immediately recognized Carmine DiGuiseppe as he flipped his cigarette butt into the street. He made his way to the door where his friends waited. The guy with the shaved head answered the door and greeted Carmine with a handshake and snap of the fingers. He introduced his cronies one by one, and they entered the house and the door shut.

"We're in. I think we should call for some backup. Let's wait, because I don't feel like busting in there and kicking everyone's ass without you and the rest of the guys enjoying some fun."

"Monsternick, you are too kind. I would love to see that. By the way, we're pretty close to ten thirty-two, so you owe me forty bucks."

"You stinker. Since you're new, I'll throw in the extra twenty. By the time the guys get here, the kids should be quite involved in their game. We'll scope out the scene and have everyone close by for quick infiltration. They won't know what hit them. I feel like we should be playing the theme song from *Cops*. Better yet, I wish we had a camera crew to see the look on Carmine's face when we meet up with him again."

Inside the house the beer was flowing and a haze of smoke lingered in the room. Belches and farts filled the air as the boys anted up and raised each other with marginal hands. None of them were well versed in poker, but Carmine and Jo Jo were slowly gaining their confidence that it was all fun and games. They would deliberately lose hands while the young guns would wipe away the chips on the table with wide grins. Little did they know their paychecks were soon to be donated to Jo Jo and Carmine as they hustled the boys while the alcohol permeated their brain cells.

"Yo bitch, I raise you ten bucks," said Tyrone, blowing cigar smoke in the air.

"I see that and raise another five," replied Carmine.

It was just the two of them at this point. Everyone else had folded their cards. Once in a while on these Friday-night gigs, Carmine would get into a pissing contest with some of his buddies. They would raise each other until the pot was well over a hundred dollars, and then one of them would fold. It was usually Carmine, because at this point he was still feeding the egos of his half-baked friends for the evening and needed to pump them up.

After four more rounds of raising bets, Carmine decided to call his nemesis for the round. Tyrone laid down two pair aces high. "Beat that, sucker," called Tyrone.

Carmine had a full house with kings, but he feigned defeat. "You win, Ty. I can't beat that. I just wanted you to give in. You know, once you start these rolls, you can't stop, but this time I'm gonna let you win."

Throughout the next hour, Jo Jo and Carmine slowly began their uprising against the amateur card sharks sitting at the table. Jo Jo was also continually opening fresh beers for the now inebriated players. Their speech was slurred and all of them were sitting with cigarettes or cigars hanging out of their mouths. Not one hand was finished without someone going to the bathroom. With all of these distractions, Jo Jo and Carmine had a chance to slip in some of the cards they had tucked underneath their legs. No one ever knew because the alcohol had slowed all of their senses. Jo Jo and Carmine claimed to be drinking vodka and cranberry juice, but it was really just cranberry juice. They were as sober as an alcoholic coming out of the Betty Ford Clinic. They remained sharp as tacks while the surrounding party was slowly degenerating into blithering idiots.

"Yo Carmine," started Jeff, who was a newcomer to the group. "What do you know about the World Cup? Can I get in on some of the action? I want Germany to kick the snot out of everybody. That's where my ancestors are from and I truly believe they'll be the champions this year."

His slurred words brought a smile to Carmine's face. "Well, buddy, let me tell you. There's some fantastic money to be made on the World Cup. Not many people bet on that stuff, but if you put enough down, you can make a load of money. What are you thinking?"

"I don't know, maybe a couple hundred bucks."

"Jeff, if you can come up with a grand, I can guarantee that if you win, you'll come out with at least five. Germany isn't favored to win, and their odds are pretty slim. Do you think you could get me some money by Monday? The Cup starts on Tuesday, you know. I need the money before it starts or the bets won't count."

"I'll do everything I can, man. Hey, Ty, why don't you lend me a few hundred bucks? You have been cleaning house here tonight."

"Yeah, bro, if I keep winning like this, I will gladly lend you a few hundred. You have to pay me back with interest, though. Let's say a case of Heineken for every hundred."

"You're on, Ty. Carmine, I'll have the money for you ASAP."

Jo Jo interrupted the conversation. "All right, enough. Whose turn is it to deal? I'm ready for another hand."

Carmine picked up the cards and began to shuffle them. After he finished, he lit a cigarette and took a long drag. He blew out some smoke rings and they hovered over the table. "This round, we're playing seven-card draw. One-eyed jacks are wild and so are deuces. I have a feeling this is where my luck is going to change. I haven't won a hand all night. It's time for the table to turn."

That was Carmine's way of sending a message to Jo Jo. It was time for the cheating to begin. Their winning hands were tucked away underneath their legs as Carmine dealt the cards. The intoxicated boys sat there staring awkwardly at their hands. Jo Jo and Carmine could instantly tell by the looks on their faces who had decent hands and who did not. Carmine and Jo Jo started out the betting with a minimum number. Slowly but surely, the pot began to build. The cronies in cahoots were nudging each other under the table to pass strategies to beat their opponents. Jo Jo had enticed two of the kids to relinquish a significant amount of their money into the pot. He pretended nervous gestures, and of course they were so obvious, the drunken boys picked up on them. They decided to raise the bets. Jo Jo reluctantly agreed and laid down his chips. "Well," said Jo Jo. "I can't raise anymore. What do you have?"

Tyrone laid out a full house king high. Jo Jo had four jacks. The boys had no idea that the other players had the jacks, so Jo Jo laid them on the table and announced his hand. "I have four jacks, son."

At that moment the door bust open like a shotgun blasting into a flock of birds. "Everyone get down! Get down now!" screamed Melinda as she rushed to the table with her gun drawn. The boys floundered for a moment and made their way to the floor. "This is a bust. We have a warrant to search the premises."

"What do you mean?" said Jo Jo. "We're just playing some cards. We haven't done anything wrong. We're just a bunch of dudes hanging out on a Friday night."

"Shut up, jackass," yelled Melinda. "We have reason to believe that a lot of drugs and money have been coming through this house. We need everyone out while we investigate." She looked over at Carmine. The warrant was bogus, but they needed a reason to get Carmine into custody. "Say," gestured Melinda in a sarcastic tone, "aren't you Carmine DiGuiseppe?"

"No," he replied. "I don't know what you are talking about. Who's that?"

"Don't give me your bullshit," interjected Fredric. "Let's see some ID."

Carmine pulled out his wallet from his back pocket with two identifications. One had him registered as a Fernando Cortez, while the other identified him as Carmine DiGuiseppe. By this point, the other boys on the floor had their hands behind their heads and had pretty much soiled themselves. They'd been expecting a great night of drinking and playing cards. The last thought in their mind was to end up laying on the ground waiting to be cuffed and stuffed.

The backup crew did a sweep of the house that turned up nothing of significance. The only item of importance was the fully stocked fridge. The boys were identified as underage, and the intoxication proved their guilt. "Who purchased the alcohol?" inquired Melinda.

"It was here when we arrived," blurted out Jeff. This was his first time attending the Friday-night card games, and his first run-in with the police. He was ready to sing like a canary, for he had an almost fully paid scholarship to Villanova that he was not about to give up.

"So, are you saying that this man here with the bald head has provided all of the alcohol for you this evening?"

"Yes, ma'am, that's true," answered Jeff.

Jo Jo shot the look of death at Jeff, who could not have cared less at this point. He was scared sober and nothing was going to stop him from passing the buck. Melinda looked over at Jo Jo. "Is that true, sir?"

"He's lying," said Jo Jo. "These guys brought the alcohol with them. I don't know what he's talking about."

"Fredric, go into the kitchen and look in the garbage can. Look for any receipts that would prove purchase of alcohol."

Fredric went into the kitchen that was now littered with beer bottles. He dug through the trash of bottle caps and empty potato chip bags until he found a brown paper bag. As he searched inside, he found a receipt for several six packs that had been paid for with a credit card owned by a Joseph Nebula. He brought the receipt back out to the living room where all of the hoodlums were lying on the ground with their hands tied behind their backs with plastic cuff restraints. "Monsternick, it says here that a Mister Joseph Nebula purchased several six packs with his credit card tonight. I haven't found any other receipts in the garbage can. It looks like our friend here has been fibbing to us. Have you been fibbing to us, Mr. Nebula?" Fredric leaned down and gave Jo Jo a noogie on the head. "You wouldn't do that to us now, would you?"

Jo Jo grunted and said nothing. "We need to get these guys to the station so we can clear this up," said Monsternick. "Jo Jo, we found no evidence of drugs in this house except for a small amount of marijuana that was probably for personal consumption. We were misinformed about the drugs running through this house, but you are serving alcohol to minors and gambling illegally. That's just as bad. We'll have to take a ride and get our stories straight, boys. I hope you have your get-out-of-jail-free cards. Your parents probably won't be happy with you tonight. It looks as if you'll all be folding for the evening. Hey, Hassie, do you remember that Kenny Rogers song about the gambler? 'You gotta know when to hold 'em, know when to fold 'em, know when to walk away, and know when to run...'" Melinda laughed and signaled the other officers to pack the boys up and take them to the station.

❧   ❧

Back at the station, everyone had been booked and smiled pretty for the camera. Fingerprints were taken, and then the boys were led to a holding cell until their parents came. Carmine and Jo Jo were placed in separate interrogation rooms.

Jo Jo was the first to go under the light. Melinda pulled up a chair and looked him straight in the eye. He took a deep breath through his nose and smirked. "What's this all about? I have no record and I've never been arrested. For an offense like this, first time will be a slap on the wrist and a minimal fine. Why did you come busting into my house?"

"Jo Jo, I'll ask the questions. You give the answers. Your boy Carmine, how long have you two been running together?"

"I've known him for a few years now. He's not that bad a kid. My brother went to school with him. Then he went into county."

"Are you guys running bets?" asked Melinda.

"What do you mean?" asked Jo Jo disgustedly. "This is ridiculous! I don't need to listen to any of this. I want my lawyer."

Melinda stood up and threw the ashtray against the wall where it shattered into a hundred pieces. "Listen, you bastard, just because I'm a girl doesn't mean I can't be a raving bitch. You give me answers or I'll press charges against you, and trust me, the DA will make sure you see some time with the big boys. You ever been to prison, Jo Jo? Well, let me tell you, I don't think you'd like wearing a pink tutu and dropping the soap in the shower, if you know what I mean."

Jo Jo sat in shock. He had not expected this middle-aged, blond-haired cutie to be ripping into him. Surprisingly for someone of Jo Jo's stature and appearance, he changed his tone. "All right, settle down," he yelled. "Look, I don't run bets with him. He does that on his own. We just play cards together. He brings his buddies up from school and we hustle them a bit. Nobody gets hurt and it's a little fun. Come on, you have better things to do with your time than bust me for playing some cards with high school kids. Christ, most of them probably have their own fake IDs anyway."

"Jo Jo, let me tell you something. What you're doing is illegal, and you should go to jail and experience what it's really like to be someone's beach ball. However, in light of your cooperation, I'm going to let you off with some fines. If I ever hear your name associated with poker, alcohol, high school kids, or Carmine DiGuiseppe, I promise you'll be in the slammer and I'll make sure to tell everyone that you are a child-molesting pedophile. You know why?"

Jo Jo scrunched up his face. "What the hell are you talking about? I don't do that stuff."

"I know, but when those guys go to jail, they get it the worst, if you catch my drift. So, you keep your nose clean and we won't bother you. We have you on file, too, so don't think we won't be notifying our guys up there in the Northeast to keep an eye on you. I can make it all happen with a snap of the fingers."

Melinda snapped her fingers and folded her arms. The seriousness in her face was enough to persuade Jo Jo that he was going to be making some changes in his lifestyle over the next few days. Bad enough his brother was in prison, and Jo Jo had already heard the horror stories from him. Melinda slammed the door behind her and entered interrogation room B where Carmine sat with his legs crossed and a cigarette hanging from his mouth.

"Hello, ma'am, you're Detective Monsternick, right?" he asked politely. "I am so sorry about tonight. I wasn't drinking anything, I swear. You can give me a field sobriety test, I will pee in a cup, donate blood, whatever you need. We were just having fun."

"Mr. DiGuiseppe, I respect your sincerity, but according to your compadre next door, you bring your supposed friends up there every Friday night to get plastered and donate money to your scholarship fund."

"Well, I'm planning on going to college next year. I have great grades and I've been accepted to Drexel University. I plan on studying chemical engineering."

"Carmine, I've spoken to Principal Monk on several occasions. You won't be graduating this year, and if you do attend high school one more year, you'll be old enough to buy alcohol. Your charisma and flaky personality won't float here. I know who you are, and there is no way you could convince me otherwise."

"Please, Detective, I think you have me confused with someone else. I do very well in school. Principal Monk just doesn't understand me. I want to do well, but I get into trouble because my parents don't pay attention to me. So when I get in trouble, they have to acknowledge me. Any attention is good attention to me." Carmine's eyes welled up with tears. Melinda laughed as she walked out the door. She went into the room where Chief Cahill and Fredric were standing. "Do you think that's really the truth?" asked Fredric, who had been listening to the interrogation.

"Hassie, this kid is about as smooth as a gravel driveway. He may have a lousy home life, but not lousy enough to fool me into believing him. You'll learn. Go get the Drater Club notebook and we'll shake him up. Hassie, watch how his demeanor changes when I throw it on the table. I guarantee he gives us something. Watch and learn."

She came back in with the evidence tucked in a folder. She pulled up a chair and offered Carmine some tissues. "Look, I'm sorry, son. I know you have a lot to deal with in life, but it doesn't mean you have to do bad things for attention. You can do positive things and the rewards will pay off handsomely. I just need to ask you a few things. Did you know either Adrian Belkoy or Phillipe Mindrago?"

"I only knew them through school," he said between sobs. "It was horrible what happened to them. I was so scared when that happened."

"Why was that, Carmine? Were you afraid that you were next?"

Carmine stopped crying and looked directly at Melinda. "What do you mean?"

"I mean, were you afraid you were next because they were running all of their bets through you and making loads of money?"

Carmine straightened up in his chair. "How do you know about that?"

"First of all, Jo Jo already confessed because I threatened some serious jail time and reforming of his sphincter, and second of all, your name is in this book." She tossed the Drater Club notebook on the table. Carmine stared at it blankly, looking like the wind had just been taken out of his sails.

"Where did you find that?" he asked.

"In Adrian's locker. Start talking."

Carmine's sobs had turned to short breaths that were reminiscent of a nervous schoolboy in gym class who hyperventilated when it came time to run the mile. Melinda knew she had him and turned to the two-way mirror and winked. Fredric and Chief Cahill stood there laughing. Fredric was now convinced that his master was the true oracle of knowledge.

Carmine lit a cigarette and took a long drag. After slowly exhaling and letting the smoke curl up around his nose, he began to speak. "Well, I've been running bets for a while now. I don't do anything big like over a grand or two. It's a great way to make money, and I know college is out of the question. I won't ever graduate high school, so I need a future. I've stayed under the radar for this long; I figure I can get by without getting caught."

"So what does the Drater Club stand for?" questioned Melinda.

"It's 'retard' spelled backwards. Adrian knew he was a numbers whiz, so he thought everybody else was a retard, even the teachers. He just kissed their ass for high grades. That was one smart kid."

"So, you took all of their bets for them?"

"Yeah."

"What if they wanted to go over two grand?"

"I could cover them," he said. "We had a rolling system going where they weren't ready to cash out, so I set up an account for them."

"Why did you run from us that night in Manayunk?"

"I don't know. You guys wigged me out. I thought you were coming after me."

"You mean like we were bad guys, Carmine?"

"I don't know," he stammered. His tone changed to one of belligerence. "I just got scared. That's all."

Melinda could tell he was holding something back. "What about the gun?"

"I carry it for personal protection."

"Have you ever been harassed by the bigger bookie guys in town?"

Carmine stiffened a little in his chair. He cleared his throat and his eyes shifted around the room. He would not tell her about Vinnie and Jimmy, because if the police knew about his involvement with the mob, he would be dead for sure. Instead, he made up a story about how he never had any encounters with them because the people they dealt with were ones not in school and business people. "No, those guys are for adults. I only work with kids."

Melinda could tell he was scared, but she wasn't sure why. Carmine knew it would be safer to take his chances instead of ratting out the mob and risking witness protection. The last guy who did that woke up with rigor mortis.

"Well, you'll be taken to a holding cell until you make bail. I'll see you in court, buddy. And If I were you, I'd be very careful on the streets. You never know who might be watching you, Carmine." Melinda smiled and walked out the door.

Back in the other room Fredric was waiting. "Monsternick, that was amazing. I think I smell a Tony."

"Yeah, but he's not telling us everything. He knows something else and he's holding out. We'll have to keep an eye out on him. He is one sneaky little sot."

∽ ∾

Vocabulary

BAC   Blood alcohol content.

Breathalyzer test   A device used by law enforcement to acquire the blood alcohol content of suspected intoxicated individuals.

DWI   Driving while intoxicated.

Ethyl alcohol   The type of alcohol produced by yeast fermentation of sugar under ordinary conditions. It is also referred to as ethanol, the alcohol in beverages.

Excretion   The loss of ethanol from the body through urine, sweat, breath, and other routes of exit.

GCMS (gas chromatography coupled with mass spectrometry)   A two-part process involving the separation of a chemical mixture into pure chemicals followed by the identification and quantity of the chemicals.

Henry's Law A law proposed by physician William Henry in 1800 that states "The amount of any given gas that will dissolve in a liquid at a given temperature is a function of the partial pressure of that gas in contact with the liquid and the solubility coefficient of the gas in the particular liquid."

Horizontal gaze nystagmus An involuntary jerking of the eye that occurs as a person follows a stimulus with the eyes to one side or another. If a person is under the influence of alcohol or certain drugs, the jerking will tend to become distinct when the eye reaches a certain angle in reference to the stimulus.

Isopropyl alcohol An artificial fatty alcohol used as an antiseptic or solvent. Also called rubbing alcohol.

Methyl alcohol A light, volatile, flammable poisonous liquid alcohol used as an antifreeze, solvent, fuel, and denaturant for ethyl alcohol. Sometimes referred to as wood alcohol.

Miranda rights Informs a person that he or she has the right to remain silent and the right to an attorney. They must be read before a person is placed under arrest.

Oxidation The addition of oxygen, removal of hydrogen, or removal of electrons from an element or compound. In the environment, organic matter is oxidized to more stable substances. The opposite of reduction.

Background Information

There are a number of different alcohols, such as methyl alcohol (wood), isopropyl alcohol (rubbing), and ethyl alcohol (ethanol). The last alcohol is found in the beverages consumed by the public. All of these alcohols are volatile and have a low boiling point. They are also colorless and practically odorless. When someone is suspected of intoxication, a police officer may notice the smell of alcohol on the person. This is not the alcohol itself, but rather the ingredients used in conjunction with the alcohol. For example, the malt, hops, and barley found in beer gives the smell we associate with someone who has been drinking beer. It works the same way with other types of wine and liquor.

Alcohol has the greatest effect on the central nervous system. It acts like a depressant and is directly proportional to the concentration of alcohol within the nerve cells. As more alcohol is consumed, the functions of the central and rear portions of the brain become affected .

When alcohol reaches the stomach, it is not digested. Some of it is absorbed into the bloodstream through the stomach walls, while the remaining alcohol is absorbed through the small intestine.

There are many factors that determine the rate at which the alcohol is absorbed into the bloodstream, including how long it takes a person to consume the alcohol, the percentage of alcohol in the drink, the amount consumed, the type of food eaten during the time of consumption and how much is present in the stomach, the size of the person, and the person's physiological makeup. The peak of absorption usually occurs within 30 to 90 minutes after the last drink has been finished.

As the alcohol is circulated through the bloodstream, the body immediately begins to break it down through oxidation and excretion. Most of the alcohol consumed will be oxidized to carbon dioxide and water. The organ central to this process is the liver. An enzyme called alcohol dehydrogenase converts the alcohol into acetaldehyde and then into acetic acid. In turn, the acetic acid is oxidized once more to carbon dioxide and water.

The remaining alcohol is excreted unchanged in urine, breath, and perspiration. One of the more significant findings through extensive testing is that the amount of alcohol exhaled from the lungs is directly proportional to the concentration in the blood. This led to the development of the Breathalyzer (discussed in the following section). As blood travels throughout the body, the capillaries connect the arteries with the veins. It is here that the exchange of materials between the blood and other tissues take place. It is in the lungs that the respiratory system bridges with the circulatory system so oxygen can enter the blood as the carbon dioxide leaves. The capillaries in the lungs are in close proximity to the alveoli, which are located at the ends of the bronchial tubes. Within the lungs, there are approximately 250 million alveoli. The tubes themselves connect to the trachea, which leads up to the mouth and nose where breath is expelled.

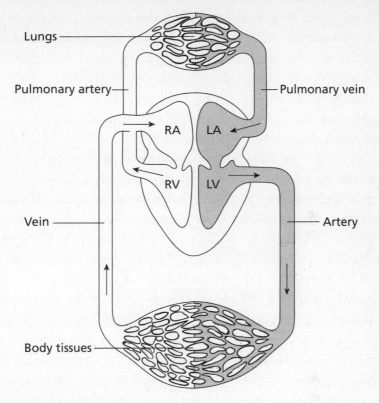

Lungs

Pulmonary artery

Pulmonary vein

RA   LA

RV   LV

Vein

Artery

Body tissues

The human circulatory system.

Now consider how alcohol becomes part of your expelled breath. As the blood flowing through the capillaries comes in contact with surface of the alveolar sacs, an exchange takes place between the fresh air in the sacs and the deoxygenated blood in the veins. The oxygen passes through the walls of the alveoli into the blood while the carbon dioxide is released into the air. If alcohol is present in the body, it will also pass and be expelled through the nose and mouth. It is here that the breath can be used for identification of blood alcohol content, or BAC. One can use Henry's Law as a good explanation as to how the alcohol becomes present in breath. William Henry was a physician who eventually dedicated his time to studying chemicals, with a special interest in gases. Henry's Law states that when a volatile chemical (alcohol) is dissolved in a liquid (blood) and is brought to equilibrium with air (alveolar breath), there is a fixed ratio between the concentration of the volatile compound in the air and the concentration in the liquid and this ratio is constant for a given temperature. Therefore, when one considers that the temperature of air coming from the mouth is normally 34 degrees Celsius, the ratio of alcohol in the blood to alcohol in alveoli air is approximately 2,100 to 1. This converts as 1 milliliter of blood will contain nearly the same amount of alcohol in 2,100 milliliters of alveolar breath.

As for the rest of the body, the blood circulates to other areas and organs. As the arteries branch out into capillaries, the alcohol moves into the tissues. The alcohol has a higher propensity to move into tissues and organs that are more watery. Fat, bones, and hair are low in water content and little alcohol will be found there.

Forensic Implications and Jurisprudence

The law enforcement officer must follow a protocol when someone is under suspicion of intoxication. First, the officer may smell what seems to be alcohol. If this is the case, the officer will usually conduct a series of field sobriety tests. Another indicator might be slurred speech, or the person might seem out of sorts—slow to respond to a question, delirious, defensive, or belligerent. In some states, a Breathalyzer test is not permitted at the scene, so field tests are done to determine if further testing is necessary. These tests consist of a series of psychophysical tests, including the horizontal gaze nystagmus, walk-and-turn, and the one-leg stand. In states where the field Breathalyzer is permitted, it may be used at the scene to ascertain the individual's blood alcohol content.

Horizontal gaze nystagmus is an involuntary jerking of the eye as it moves to the side. The intoxicated person is unaware it is happening and does not have any control over the jerking motion. The subject is asked to follow an object such as a penlight or finger. If the latter is used, a flashlight will accompany the object. As the suspect's eyes shift to the side following the pen, the eye will involuntarily jerk back the other way, indicating the nystagmus. The more intoxicated the person is, the less distance the pen has to move before the eye jerks back. If the subject is under the influence of drugs such as barbiturates or other depressants, the nystagmus might appear sooner.

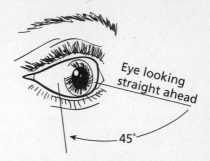

Horizontal gaze nystagmus.

The walk-and-turn and one-leg stand both test the person's ability to comprehend and execute instructions simultaneously. In the walk-and-turn test, the subject must walk a straight line, keeping his or her balance while going heel to toe. The one-leg test requires the subject to maintain balance while standing on one leg lifted several inches off the ground while reciting the alphabet or counting. Every police department is different and usually has its own prescribed rules for possible DWIs.

These tests are not exclusive to drivers on the road. They may also be used for anyone suspected of intoxication to the point where it may be hazardous to themselves or other people. One situation where it is often used is when an underage drinking party is broken up by the police. These tests will indicate if the person should be tested further, resulting in a possible legal offense.

The Breathalyzer was developed as a noninvasive method to test suspects at the scene or police station. It was first developed by R.F. Borkenstein in 1954. It initially was a system that collected a sample of alveolar breath and placed it in a heated cylinder. When the instrument was placed into the analyze position, the breath sample passed onto a glass ampoule that contained potassium dichromate and silver nitrate in sulfuric acid and water. If alcohol were present, it would be immediately dissolved in the dichromate solution and oxidized to acetic acid. The amount of potassium destroyed would be proportional to the amount of alcohol passed into the ampoule. The Breathalyzer acted as a spectrophotometer in the respect that the absorption of light passing through the potassium dichromate was measured at a single wavelength.

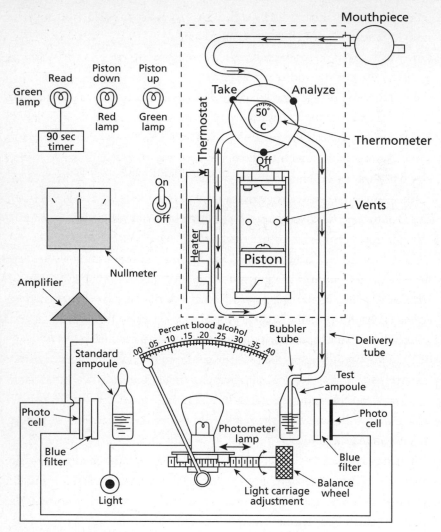

The original Breathalyzer.

The Breathalyzer has undergone several modifications and police now have a field Breathalyzer, which is approximately the size of a pocket calculator and uses a fuel cell to compute the estimated blood alcohol content (BAC). After the suspect has blown a sample into the tube, the fuel cell absorbs the breath sample, oxidizes it, and produces an electrical current proportional to the BAC. The fuel cell itself will last approximately 3 to 5 years before it must be replaced. Of course, calibration is essential, for tests may be rendered inconclusive in court if not properly calibrated.

Newer technology has developed a passive alcohol screening device, which is used to sample the air around the person, including alveolar breath. It is considered passive because it does not require the suspect to perform an act such as give blood or urine or blow into a tube. It is called the Lion Passive Alcometer and is approximately the size of a pen flashlight. It incorporates a switch that draws in the breath and another that processes the sample. It takes approximately 10 seconds for a reading. It is used as a tool for probable cause and might not necessarily be admissible in court.

The standard for intoxication used to be .10 percent, but has been reduced to .08 percent in most states, while others have gone as low as .05 percent. If the suspect meets the requirements to be considered intoxicated, a blood or urine test is necessary to confirm the BAC. In this instance, suspects will be taken to the hospital or certified official.

In 1973, the Highway Traffic Safety Administration created a statute that states that the operation of a motor vehicle on a public highway automatically carries with it the stipulation that the driver will be given the choice of either submitting a specimen for an intoxication test, or possibly be arrested on the spot. If the driver decides to supply a specimen, it is called "implied consent." If the driver refuses the test, he or she may be subject to loss of license for a period that usually covers six months to a year for the first offense. In cases of repeat offenders, the license may be revoked on the spot.

When a suspect fails the Breathalyzer test—meaning the suspect has a BAC higher than the legal limit—and/or other tests, he or she is arrested for a DWI. Miranda rights must first be read in order to establish the suspect's legal rights to speak or wait until an attorney is present. As mentioned earlier, the right to not submit to a test is a separate issue, therefore delineating different consequences. When a person is in an intoxicated state, he or she might not understand the difference between the two and refuse the submission of a specimen under the Miranda rights. The Miranda rights do not protect the driver from having to submit a specimen or from possibly losing his or her driver's license.

In some states a Breathalyzer is admissible in court, in others it is not. For instances where samples are necessary, they must be taken under conditions that are most conducive to valid results. There are environmental and biological factors that may affect a sample. For example, if a urine sample is left open, contaminants can oxidize the sample and provide an incorrect reading. Blood should be stored in a refrigerated area to prevent spoilage. When the blood sample is taken, a nonalcoholic disinfectant should be applied; however, some argue that this may affect the outcome of the test. Once the area is sterile, the blood is collected and placed in an airtight container, and an anticoagulant and preservative are added. One common anticoagulant is EDTA (ethylenediaminetetraacetic acid) and a common preservative is sodium fluoride. The anticoagulant prevents the blood from clotting while the preservative inhibits the development of microorganisms that can destroy the alcohol.

Once the sample is taken, the most common and effective method of instrumentation is use of gas chromatography coupled with mass spectrometry (GCMS). When the GCMS is used correctly, the alcohol is separated from other volatiles that may exist in the blood. By comparing the results with known standards of alcohol, the scientist can determine the alcohol level with a high degree of accuracy and scientific validity.

The experts who testify in a DWI case may range from police officers to scientists. When the Breathalyzer is considered the chemical instrumentation for acquiring BAC, the witness need not be a chemist or other type of scientist. As long as the instrument is calibrated correctly,

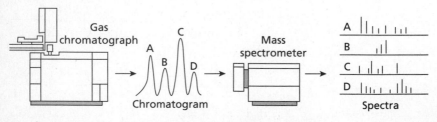

Gas chromatography coupled with mass spectrometry (GCMS).

it will process the sample and produce a digital display and possible printout of the results. However, it is paramount that the person administering the test be properly trained on the procedures that pertain to handling of the instrument and the conditions in which the test should be conducted.

In some jurisdictions, only those who have passed state board examinations are considered qualified experts. Of course, where other tests are performed that involve the assistance of a certified scientist, the accompanying credentials must be established.

11

The Gators
Get Served

Before Adrian and Phillipe were killed, they spent Monday evenings at the Turf Club in the city. In a room full of dedicated drinkers and smokers, they carefully read through the program of upcoming races for the evening. Phillipe learned about horse racing from his father, who had once invested some money with a partner on a thoroughbred racehorse named Smarty Pants. Smarty eventually became a world-renowned racehorse and the icon of horse racing. His last year led him through the ranks toward the Triple Crown and landed him in the Belmont Stakes, where he broke his leg in the last turn. More than a few people were devastated, because Smarty had been a local hero in the city of Brotherly Love and a large number of people lost their shirts.

Phillipe was well versed in examining track conditions and figuring out who liked the slop and who preferred dry grass. Adrian played his part by crunching numbers and figuring out their best odds and how to make the most of their money. Since the two were over 18, they could collect their winnings, which usually did not amount to more than a few hundred bucks.

As they sat there munching on nachos, sipping sodas, and waiting for the next race, Adrian broke the silence. "Hey, bro, remember that Superbowl gig? That was one killer party we went to with Gizep. That dude knows how to do it right."

"Yes, he does," replied Phillipe. "Those lobster tails and petite filets were solid, man. Listen, how are we doing with our March Madness pools?"

"Right now, I'm in front of you by two teams. I don't know how anyone else is doing, but I have a spotless bracket. How much have you made so far in the individual bets with Gizep?"

"All together, probably over two Gs. My prediction is that Villanova and Florida will be in the final. The Gators are on fire this year and the Wildcats might stand a chance if Johnson recovers from his sprained ankle. It should be a great matchup."

"You haven't cashed out yet, have you?" asked Adrian.

"No. What about you?"

"Nope. In fact, I have close to five grand stored up with Carmine. I have an idea, Phillipe. Why don't we combine our winnings and make one huge bet on the final game?"

"What if we lose? I don't have the cash to cover that, do you?"

"Are you kidding me?" joked Adrian. "We've been on a roll, man! You won't need any money because we won't be losing. We could make a load and throw a bash! We just need to wait and see who's in the final game, and I'll crunch the stats for us. I can't believe you would even fathom the idea of losing. It's not in my vocabulary right now. Besides, I haven't steered you wrong yet, have I, bro?"

"No, you have not, my man. The Drater Club shall live in infamy at Roxborough High. I don't think anyone will ever compare to us."

"That's right, buddy," nodded Adrian. "They're just a bunch of retards. Especially Monk. That guy is so easy to win over. All you have to do is talk sports, academics, and future in college, and you're in. The other retard there is Simons. He's afraid of his own shadow. That guy hasn't taught us a lick this year. All he does is sit there behind his desk and read the newspaper online. God forbid you ask him a question about anything other than stats. He shrivels in fear like the Hunchback of Notre Dame. Not to mention Notre Dame will be kicking butt next year under the direction of Charlie Weiss. I can't wait to get to college next year. So many opportunities for us young strapping lads, right?"

"You betcha," replied Phillipe.

"Hey, man, you'll keep this gig going when you get to college, right?"

"Does a bear shit in the woods? Is a frog's ass watertight? Of course I will," said Phillipe. "You will, too. How can you give up the free cash?"

"Exactly," agreed Adrian. "We've done so well and I can only win more money. We need to stay in touch, too. Let's make a pact right now that no matter what, the Drater Club will never die. Even when we get old and have kids and stuff, we'll still be bringing in the money. It's just too easy."

"Right on, man," Phillipe agreed as he slapped Adrian on the back. "I can see it now. A big ass Escalade and Porsche in the driveway, huge mansion, hot maids from Europe who walk around in skimpy clothing, awesome game room, and a pool with a bubbling Jacuzzi. Ahh, life is good."

With that the next race started and their dreams were filed in the depths of their minds. As far as they were concerned, they were invincible and nothing could stop them, the dynamic duo of Adrian and Phillipe keeping their secret society to themselves like a private fraternity for non-draters.

The race was over in less than a minute and Adrian and Phillipe slapped high fives as their horse made them a hundred bucks.

"Do you think you'll have kids?" inquired Adrian.

"Not for a long time. I want to be like Wilt Chamberlain. That guy was the man. When I get bored with all the chicks, then I'll settle down and have some kids. They'll all be studs just like me. I can see it now. They'll be the most awesome athletes and have all the chicks. A real chip off the block, if you know what I mean."

"Of course, it wouldn't be any other way. I don't know about me," said Adrian. "I think I might want to have kids. I don't have any brothers or sisters and it was kind of lonely growing up. It would have been cool to have a brother to hang out with. You're kind of like the brother I never had, Phillipe. We're pretty tight, man. We've known each other since grammar school. The only other person I'm close with is my dad. He doesn't know everything we do, but he's a great guy to hang out with. He would probably smack me upside the head if he knew about the Drater Club, but he'll never know, will he?"

"Never," Phillipe said as he shook his head. "It's just you, me, Carmine, and a select other few."

The two of them finished an evening filled with jocularity, confessions of life long dreams, and aspirations of the future in college.

∽ ∼

The end of March quickly approached and college basketball was coming to an end. The championship game was two days away. The Villanova Wildcats and Florida Gators would meet each other in a final matchup of dexterity, endurance, and obsession for a national title.

Adrian put in some serious time studying his notes from the season. He had all of the statistics printed out for both teams. His proficiency in the laws of probability was sure-fire, and there was not one shred of doubt in his mind that he could successfully pick the winning team. After he compiled all of his notes and weighed the pros and cons, he called Phillipe. "Yo, what's kickin', chicken?"

"Whatta you got, Adrian? Wait, don't tell me. I've been reading ESPN's Web site for the last hour. It sounds to me like Florida is the winner. The Wildcats are down a starting forward and guard. There's no way they'll be able to contain the monster starting lineup of the Gators."

"Well, my friend," replied Adrian, "I couldn't have said it better myself. The Gators are going to dominate the game. I've worked out all the numbers and our chances of losing are pretty much slim to nil. I say we combine our money and throw it all down on the Gators."

"Bro, by this point in the season, we stand a chance to win, like, fifteen grand! You should call Gizep and make sure we get the bet in. I would imagine he's going to be taking on a lot of action."

"Right," said Adrian, "I'll call him now and I'll talk to you at school tomorrow. Hey, when we get to Simons' class at the end of the day, let's see if we can run our theory by him about the game and see what he says.

He hates getting into those conversations because he has to pick a side. Too bad he's a big wuss and probably will start cowering in his shoes because he might upset someone. Later, tater."

Adrian hung up and double-checked his notes. He was about to embark on the bet of a lifetime. Although probability looks at possibilities, it is not definitive, just like anything else in life. But for him, he was the smartest kid in the world, and could never be wrong about his prediction. He made it this far, so why worry now? Being a popular athlete and scholar floated his boat and his ego grew two sizes. He picked up the phone and called his trusty old pal Carmine.

"Yo Gizep, what up?"

"Nothing. Just chillin'. Whatta you want for the game? I'm assuming you're taking the Gators?"

"I wouldn't have it any other way, bro. There is no way they can lose. Nova is down two studs and the second string for Florida can whip the first string. It's a no-brainer, man. Say, Phillipe and me want to know if we can put our money together and make one big-ass bet. We haven't cashed anything out, so you should have the money to cover that, right?"

"Dude, what do you take me for, a redneck out bangin' sheep on a Friday night? Of course I got you. Haven't I taken care of you all season? You know I have your back. I'll take care of it and I expect you and Phillipe to join me for the game. I'm going to a friend's house and both of you are invited. Don't bring anything except your checkbook in case you lose. It's just business."

"You wise ass. We haven't had to pay you yet, you son of a building block. Where should we meet you?"

"Meet me down by Hikaru and we can drive up together. I think you'll have a fantastic time."

"Right on, bro. We'll see you then."

Adrian hung up and lay in his bed trying to fall asleep. It was difficult because he kept playing out the game through his head and walking away with a stack of cash to spend. Then he began to think of ways he could liquidate the cash without anything to show for it. The best idea he could come up with was a huge party somewhere downtown with lots of friends, alcohol, and whatever other trinkets people wanted to bring to keep themselves occupied for the evening.

When game day rolled around Adrian and Phillipe could hardly control their emotions. They had been text messaging each other all day bragging about the winnings that were theirs after the game. Phillipe promised a private party with exclusive guests, some of which could be hired for the evening. It was a guaranteed bet and only a matter of time before they could collect.

Later they met Carmine and headed over to a house on Ridge Pike. When they walked through the doorway, they could hear the pre-game show blaring down the hallway. The floor was beautifully decorated with Italian tile and on the walls hung some famous works of art that none of the boys cared to identify. As they walked by a picture of a star-filled night sky, a man came out of the kitchen to greet them. His dark hair and complexion indicated he was of Italian descent. Thick eyebrows high-lighted his deep-set piercing brown eyes. He looked like a model for *GQ* magazine.

Gizep made the introductions and the boys eagerly shook hands with their new comrade, who could probably lavish them with a party filled with beautiful women and fine wine. His name was Mike Tamasco, and his urbane demeanor was one the boys would surely like to be part of. He invited them into the kitchen for cocktails and appetizers. On the table was a spread one would expect to see at the Ritz Carlton. Plates were filled with steamed clams, mussels, scallops wrapped in bacon, and stuffed mushrooms. Next to that was a crudités table featuring a huge cut of Grana Padana cheese imported from Italy and surrounded by grapes and crackers.

The boys licked their chops as they loaded their plates with scrumptious bites for a pre-game meal. Mike offered some wine as he stood behind a huge bar on the other side of the kitchen. The room itself was about 300 square feet. Carmine walked over to have a seat at the bar. "What do you have for us tonight, Mike T.?"

"Well, my friend, I have a sauvignon blanc from the Loire Valley and a pinot blanc from the Burgundy region. Either one will be fantastic on the palette with these appetizers. When we eat dinner, I have a '98 Bordeaux from the southern part of the region and an incredible Chianti from the Piedmonts. Boys, if you haven't noticed, I know a lot about wine; it's my living. I import grapes and juice from around the world, I sell the best products, and my customers make superior wine. I stake my reputation on it."

"Mike, these guys wouldn't know a bottle of this stuff from a Diet Coke at McDonald's," said Carmine. "How about you select a winner for us?"

Mike poured a glass for everyone, and not knowing what to look for, the uneducated boys commented how excellent the wine tasted. Mike started rattling off terms like "lemongrass" and "slight hints of vanilla with undertones of white pepper". They just nodded in agreement as they sipped the wine and stuffed their faces.

As the boys loaded themselves with as many steamers they could handle, the game started. There was some small talk before the game about players and predictions, but nothing about money. The boys were waiting for Carmine to bring something up, because they figured Mike T. was a big gambler. How else would he be able to furnish himself with such an extravagant lifestyle? Adrian and Phillipe decided not to say anything because they had the mother lode riding on this one, and they did not want to jinx their luck.

As the game started, the Gators dominated, just as Adrian predicted. They were nailing three-pointers and sinking foul shots like it was nobody's business. By halftime, the Gators were up by 20 points. Phillipe stood up to stretch and smiled greedily at Adrian. He signaled for him to go into the kitchen.

They walked over to the bar to pour themselves another glass of wine. It was of no importance that each bottle cost over $100; combined, they polished off four bottles of the Bordeaux. Adrian smiled and revealed his purple-stained teeth that made him look like a little kid who just finished his bug juice at summer camp. He leaned over to speak to Phillipe in a low voice. "Dude, we are so in. If the Gators keep up the pace, they'll crush Nova and cover the spread. We're going to be living the high

life, baby. Remember that we should spend this cash on stuff that does-
n't sit around. That's why we need to rent a penthouse suite somewhere
downtown and throw the best party ever seen at Roxborough High. In
twenty years, kids will still be talking about it. No one will ever compare
to us, bro. No one!"

"Amen to that, my man," laughed Phillipe. "Can you imagine walking out of here with a freaking wad of cash like that?"

Just then Mike T. walked into the kitchen to refill his glass. The
boys commented again on how awesome the wine was and Mike
expounded for 10 minutes about acids, ph, and grape skins. All they
could think about was their jackpot, which was only an hour or so away.
They wrapped up and made their way out to the living room for the second half.

The Gators started out strong, but somewhere along the way, they
lost their groove. They started missing three-pointers, foul shots, and
even some lay-ups. Meanwhile, Nova played the game of the century during the second half. They ran an unstoppable full court press that
knocked the Gators back on their heels. Three of their hotshots fouled
out and the Wildcats took advantage and made every basket from the
foul line. Their second half field goal percentage was above 90. The crowd
was going insane and the announcers were losing their voices from
screaming and shouting. It was a true upset in the making. The basketball gods brought fate and destiny to Nova and bestowed misfortune
upon the Gators.

Adrian and Phillipe sat dumbstruck at the events unfolding before
their eyes. Knots the size of beach balls were building in their stomachs
as they watched the Gators get served by the Wildcats. Carmine and
Mike were excited because they had placed some money on Nova. They
were slamming out high fives and snaps of the fingers while Phillipe car-
ried on like a two-year-old melting down in the supermarket because his
mother wouldn't buy him candy. By the end of the game, the color had
drained from Adrian's face, and Phillipe was drenched in sweat from
yelling obscenities at the Gators and jumping around the room like a
smacked-up cokehead.

As they sat there watching Nova bask in the glory, Carmine brought out two more glasses of wine for his now financially enslaved losers. "Guys, it's just business, you know that. It was a great season. You made a lot of cash off me, especially during football season. You knew it had to catch up with you sooner or later. The Drater Club has winners and losers. It just so happens that tonight you are the losers. I know my name has shown up on the short end of the stick, and tonight it's my turn to come out on top. This is a little overwhelming for you guys, I can see that, but it was one hell of a bet. I didn't tell you guys this, but it was a little more than I could handle, so I placed your bet through Mike T. He's part of the union of betting officials, if you know what I mean. So you actually owe him close to fifteen grand. No offense guys, it's just business."

Mike T. walked out from the kitchen with a big smile. "I couldn't help but overhear what Carmine was saying to you guys, and he's right; it's just business. I don't take personal checks, so I would imagine you'll have the cash for me tomorrow after school. I'll catch up with you somewhere along the way."

"Are you shitting me?" cried Adrian. "Where am I supposed to come up with that kind of cash?"

"Well, you've been betting all season long, right? You should always have the cash to back up your mouth. Hey, at least you have the entire day to get the money. I'll see you guys tomorrow. Carmine, here's your winnings. You have five grand in there. You know you can trust me so you don't have to count it. I have to leave and meet some friends. You can let yourselves out and lock the door behind you. The alarm will set itself. Nice to meet you, boys. It's a pleasure doing business with you."

"Carmine!" screamed Phillipe after Mike T. left, "why the hell didn't you tell us you couldn't cover the bet? Why did you lie to us? Who is this dude? Is he going to kill us? Help us out here, man. Can you lend us some money? You know we're good for it."

"Well, let's see," replied Carmine matter-of-factly. These were the moments he treasured the most. He had them by the marbles and they were defenseless, just the way he liked to see his clients. "Phillipe, you put down seven grand and Adrian, you put down seventy-five hundred. I really can't spot you guys, because I need that money to pay off my other bets. If you did win, I wouldn't have enough to pay you anyway. I figured you guys would keep rolling your profits."

"Carmine, we're completely screwed here. You have to help us, man. You owe us. We been giving you business all year long," pleaded Adrian.

"Not really," Carmine returned. "See, you never paid out, so technically, that money was floating out there in space. You've been betting through me, but you never cashed out or paid up when you lost. So, I think that means we're kind of even. I'm sorry, but there's nothing I can do for you."

"Who is this Mike Tamasco guy?" inquired Adrian. "What if we can't get the money tomorrow?"

"I think he's bluffing," Carmine answered. "I haven't known him that long, but he seems like a decent guy. I'm sure if you kiss his ass and ask for some time, he'll give it to you. Hopefully, he's an easygoing guy. Look, I gotta go. I'll catch up with you guys later. Here's some money for a cab home. We need to lock up here."

Outside, Carmine locked the door and he and the boys went their separate ways. Adrian and Phillipe didn't say much more to Carmine as they split. They hailed a cab along Ridge Pike and sat staring out the window. Adrian let out a big sigh. "How much can you come up with?"

"Christ," complained Phillipe, "I'm lucky if I can scrounge up a thousand bucks. What about you?"

"I don't know if I even have that much. I have a college fund, but I can't touch that. Besides, my parents would have my head on a platter. How are we going to come up with this money?"

"Maybe my dad can help us out," offered Phillipe. "He makes great money in the real estate business and he inherited a lot of money from my grandmother in Portugal. Let's talk to Mike T. tomorrow and see if we can get an extension."

"I just hope he's as easygoing as Carmine thinks or else we're in trouble."

‍ಌ ಌ

The next day at school, the boys were as nervous as puberty-driven teens on their first date. Adrian had diarrhea and no amount of Immodium was helping. Phillipe was in and out of the nurse's office with stomach pains and he tossed his cookies a few times. Neither one could get a handle on their nerves. At lunch they could hardly eat anything. All of their friends were joking around with them and trying to cheer them up, but nothing worked.

"What's wrong with you two?" one of the girls asked.

"I think we have a slight case of food poisoning," responded Adrian.

"Why don't you guys just go home?" she said with a bewildered look on her face.

"We'll be all right," affirmed Phillipe. "We just have to make it through three more periods."

As the day rolled on, thoughts of impending doom drove out all other thoughts. Their only plan was to confront Mike T. and ask for some extra time and hope he would be cool with it. At three o'clock, the last bell rang for the day and students stampeded down the hallway as they ran into the parking lot to drive off in their cars or get on the school bus for a long bumpy ride through Roxborough. Adrian and Phillipe met at the front doors of the building. "Here, bro," Adrian tossed his car keys to Phillipe. "Go out and start the car up. I have to hit the bathroom for one more battle with my colon."

Phillipe hesitated. "Hey, he only said he would catch up with us along the way. Do you think he's out there waiting for us?"

"I doubt it. He doesn't know what time we get out, and there's another high school in Roxborough. Besides, there are people out there in their cars and getting on the bus. I shouldn't be that long. I don't think we should go home right away. either. When I come out, we'll call Carmine and get Mike T.'s phone number. Maybe we can talk to him on the phone instead of face to face."

Phillipe scanned the parking lot for Mike T. as he walked over to Adrian's car. He was not to be found, so Phillipe sat in the car, relieved, and turned on some music. As he sat scoping out the area, 15 minutes passed and the parking lot emptied out. Adrian still had not come out.

After-school sports were taking place behind the school, so Phillipe was sitting alone in the parking lot. He told himself that he would wait three more minutes and then go in to get Adrian. Those three minutes were the longest three minutes of his life. He sat there thinking about the worst-case scenario where the two of them were tortured by Mike T. with all kinds of paraphernalia that any executioner would fancy.

His last thought was one of going into the school and finding Adrian lying dead in one of the stalls. With that, he grabbed the keys out of the ignition and started back toward the school. As he neared the doors, Mike T. appeared around the corner. Phillipe froze in his tracks as he approached. Phillipe was white as a ghost and he thought about the possibility of shining Mike's shoes with a fresh spray of puke. He swallowed hard as Mike greeted him with a friendly smile. "Hi, Buddy, remember me? Yeah, well, I was in the neighborhood and wondered if you had my money. Why don't we go over here so we can talk. I wouldn't want one us to get run over in the parking lot."

Phillipe followed Mike to the side of the building that housed a little nook. It was useless as far as the building structure was concerned, but it was a great place to create a makeshift office for Mike T. "So, what do you guys have for me?"

Phillipe hesitated. "Well, we were going to call you. We don't have the money right now and we were wondering if you could give us a few days to get the cash together. It's a lot of money, you know."

"It is a lot of money, and that's why I want to collect it. See, it's been my experience that the longer you let people go, the more excuses they come up with. I don't have time for excuses. I'm a businessman and I have a lot of other items to tend to. Nothing personal, kid."

Mike T. punched Phillipe square in the eye and he fell to the ground with a whimper. He held his eye shut as Mike T. grabbed him by the collar and threw him against the brick wall. "I hate to be mean, but you and your homie owe me fourteen and a half. I don't do payment plans, so you'll need to come up with the money soon. Since your friend isn't here, I'm giving you one more day. Next time it might be a bone that gets broken, and I'm not talking about fingers or toes. I'm talking legs and arms, if you know what I mean. I know where both of you live, so don't bother skipping town or hiding out somewhere. Carmine may be your friend, but money buys everything. Have a great day and say hi to Adrian for me."

Phillipe crouched down against the wall and started crying. He watched Mike T. disappear behind the building as he felt the hot tears running down his face. For once in his life he realized that he was not invincible and was susceptible to the roller coaster of life that was now plummeting to the ground.

Adrian walked out in the parking lot and saw Phillipe sitting against the wall with his face between his hands. "Phillipe, you all right?" he yelled, running over.

Phillipe looked up and between the tears and bloodshot eye surrounded by a puffy, reddish, blotchy hue, he continued crying and grabbed Adrian by the leg. "Don't let him kill me! I'll get the money for us. Just don't let him kill me. We never should have gotten involved with this shit. I can't believe we're in this hole, man. He's gonna kill us if we don't get that money!"

The two sat there for a few minutes while Phillipe regained his composure. Adrian was trying to figure out a way to come up with some cash to spare his life along with Phillipe's. What would be the most discreet way? Adrian sat racking his brain while Phillipe mopped his face with his shirt. The threat of death was a real one, and neither one of them wanted to check out early. There were so many great things ahead to be accomplished, and now it might all be destroyed because of the Drater Club.

Adrian got Phillipe into the car and he used some tissues to clean up his face. He focused his red eyes on Adrian. "Dude, we have to get that money. I'm gonna talk to my dad and come clean. He'll understand. He's been in situations like this before. I'll take care of it, and you can take your time paying me back. You're like my brother, man, and if I can save your life, then, well, that's what brothers do."

Adrian dropped him off and drove home. As he walked in, he noticed his mother and father sitting on the couch and his mother was crying hysterically. Adrian's first thought was that Mike T. had called and told them everything. Then again, why would he do that? It had to be something else. Adrian walked over to the couch and hugged his mother. "What is it, Mom? Why are you crying?" He looked at his dad for an answer, and he just bowed his head.

"It's your grandmother, sweetie. She died today in her apartment."

Was it a coincidence? Adrian lost a bet to a possible lunatic and his grandmother in Pittsburgh died the next day. Did Mike T. drive out there and kill her? The only thoughts running through Adrian's head were Mike T., Phillipe bailing him out, and death.

"She had a heart attack," his father said as he rubbed his wife's shoulders. "It was sudden and she never had time to call the hospital. We packed up some things and we have to drive out there for a few days. She wanted to be buried there since she grew up there. We called the school already and had you excused so you can come with us. I know how much she meant to you."

Adrian didn't know whether to smile in relief or pee his pants in fear. If he left Phillipe by himself, he might be killed, and if he stayed behind, his parents would get suspicious and he would have to tell them everything. On the other hand, if he lingered behind, he could get killed, too. If Phillipe didn't talk to his dad, they were surely on the wagon to the gallows. Adrian decided to take some time with his family to console them because he had been close to his grandmother. Even though she lived on the other side of the state, he had stayed in touch with her via e-mail and phone calls. She had visited often and spent several days at a time.

An hour later, Adrian sneaked into his bedroom to call Phillipe. Hopefully he had the chance to speak with his father and promising news was waiting. Adrian closed the door and hit his speed dial. "Hey, it's me. What's going on?"

"My dad had to leave town on a business trip," Phillipe said in a monotone voice. "He won't be back for two days. I can't tell my mom because she wouldn't get it. Would you be able to talk to your parents at all and see if they could help?"

"I have a problem too, man," groaned Adrian. "My grandmother just died and we have to go to Pittsburgh to bury her. What should we do?"

"I'm going to call my dad on his cell tonight and see if I can talk to him. Is there any way you can stay with me? I'm really shaken up about all this, man. I don't want to die. Can you tell your parents that you have a big test or something and you can't miss it?"

"They already called the school and cleared it with them," replied Adrian. "I'm stuck, buddy. Do you want to come with me? I can make up some excuse to my parents."

"No," sighed Phillipe over the phone. "You go take care of your grandmother. I know you were close to her. I'll take care of us here. I think I can pull it off. You have to promise me that this is it. No more betting, right?"

"Right."

Adrian hung up and packed some clothes for his trip. He felt guilty as hell leaving Phillipe behind, but it was the only way to keep his secret. Little did he know that when he returned from his excursion to Pittsburgh, Phillipe would be missing for 24 hours already.

12

Fredric Finds His Fortune

After Carmine walked out the door from his brief stay in his jail cell, Fredric sat at his desk thinking about everything that had transpired in the last day. He pulled out the two files for Adrian Belkoy and Phillipe Mindrago. It was a puzzling crime, because nothing was adding up. Their prime suspect had just walked out the door and Fredric was running out of leads. There had to be other places to find information that would turn into a lead or at least a remnant of one.

Fredric's lips flapped as he let out a long breath and looked at his watch, which now read 1 A.M. He flipped through the reports and documents that he had already seen a thousand times. Then he came across the Drater Club notebook. Neither he nor Melinda had had a chance to look at it in depth. They had gone looking for Jo Jo and never got around to reading the rest of it. Fredric figured there was no better time than the present, so he started carefully reading through each and every page, taking notes on his yellow legal pad. It was going to be a long night.

All of Adrian's bets for college football season were in there and he won more than he lost. This kid really had the knack for betting and seemed to do quite well. Interspersed with those bets were NFL teams. The pattern started with a Thursday-night college game and ended with Monday-night football. Fredric finally made it to the page on the

Superbowl, where Adrian had chosen the Pittsburgh Steelers to win. Next to it in the margin was a notation with an "L" and two dollar signs. Fredric knew it was a thousand dollars per sign because the amount was listed there. This seemed to be a consistent identification system used throughout the book.

After that, the bets turned to college basketball. One page detailed the full bracket of the tournament. Notes were scribbled everywhere. Fredric could decipher the terminology and figured that a lot of money was wagered on the final game. On the top of the next page were several entries for games throughout the tournament, and at the end was a big "L" and 15 dollar signs next to a sad face.

Carmine's name had surfaced on the majority of bets, and, of course, Jo Jo's poker nights, but this game did not have a name next to it. Fredric wondered if Carmine ran the bet for them. It was a significant amount of money, and a pecuniary loss like that would make Carmine a rich man. There was a motive, but no solid evidence that Carmine was guilty. Fredric shook his head in confusion and sat back in his chair. He started flipping through all of the pages, and one flickered by with a blurb of black. Fredric turned back the pages. He found it and on the inside margin in the middle was a three-word sentence: "Find Mike Tamasco."

Fredric sat up in his chair and wheeled himself over to the computer. He punched in the name of Mike Tamasco, and the list was endless. A 20-page report printed out with all kinds of information about Tamasco's ties to the Philadelphia Mafia in conjunction with Frank DeLuca.

Mike T's rap sheet started when he was 14 years old. He was busted for grand theft auto of a Mercedes in a downtown hotel parking lot, and served time in a juvenile institution for one year until he was released into the custody of an uncle. His parents had been busted for insurance fraud and sent to prison during his stay in juvie. His uncle was Johnny Merlotti, who was a notorious gangster in Philadelphia. Merlotti had also served time in jail. The list went on and on as Fredric soaked in the information. After a while, his eyes started getting heavy and the words started to blur; he was falling asleep. The last words he saw before passing out at his desk were, "FBI undercover investigation still ongoing."

Fredric fell into a deep sleep and instantly started dreaming of Mike T. and his involvement in the crime. In the dream, he was sitting with Mike T. at his table in a private room at Le Bec Fin, the finest French restaurant in Philadelphia. Mobsters were supposed to eat at Italian restaurants, but Fredric's subconscious decided to add some flair to his dream. He was sitting there eating escargot with his buddies and laughing it up. His pal Tony was telling a story about a hit he had to make on a businessman who could not pay his debt on a hockey game. (Nobody really bets on hockey games, but then again, most of our dreams are not coherent.) Tony had decided that since this guy loved hockey so much, he'd beaten him with a hockey stick and stuck him in an ice bin down at one of the fish warehouses on Packer Avenue. To top it off, he'd placed a hockey puck in his mouth like an apple in a roasted pig. The table was in tears of laughter as Mike T. sat there with his hands crossed. He cracked a smile and audibly chuckled. Fredric wondered why he was not laughing like the others. While Mike T. sipped his wine and swirled his glass on the table, he started his story.

"I gotta tell you guys this story about these two kids from Roxborough. They were high school hotshots. I think one was a football star and the other was in, like, soccer or basketball or something. Anyway, this kid Carmine, who runs bets up there, came to me and asked if I could handle a bet for them. Apparently they had been betting with this kid all year and instead of cashing out, they rolled everything over and decided to bet one lump sum on—are you ready for this?—the PGA Championship! What jackass bets on golf? Anyway, they wanted Tiger Woods to take it all. I thought that was the stupidest bet in the world, seeing Tiger's father just died and he found his wife in bed with Vijay Singh."

Fredric cut Mike T. off after his sentence. "Vijay Singh? Why would she sleep with that guy?"

"Who knows, but it only made it a worse situation for these kids. So getting back to my story, this kid Carmine tells me that they've been watching too many golf movies with Kevin Costner and think that all this negative energy will make Tiger persevere and come out on top. Well, you guys remember the tournament right? Tiger lost big time and these kids were into me for fifteen grand! I call this Carmine kid and he tells me the kids don't have the money. And I say, whatta you mean they ain't got the money? He says, you know, they don't have it. So I say, do they know you put the bet through me, Mike T.? I don't have time to wait around. I'm a businessman and when people owe me money, they have to pay up or suffer the consequences. So I tell this kid Carmine that I'll take care of it. He gives me their addresses and tells me that they play in after-school sports. I think it was lacrosse or something like that. So I sneak in there one day, you know, the locker room. It smells like feet and ass in there, and I'm dying. So I find their lockers with their names on it. One says Belkoy and the other says Mindrago. I pick their locks and leave a little note for them in there. It says, your friend Carmine had some complications covering your bet, so you owe me money. I need it tomorrow. Carmine will let you know where to meet me.

"So you know how I'm funny, right? I brought some Icy Hot and dumped it in their shorts. I want people to know I have the power to get them where it counts, and that's the family jewels."

All of the guys at the table burst out laughing, including Fredric. He started to think that Mike T. was a pretty funny guy. Too bad the FBI was listening to everything he said. He knew, but decided not to tell anyone. He looked around the table and all the mobsters had their guns sitting out on the table. Fredric looked down and his gun was sitting on the table as well. He didn't think anything of it because in this dream, he was one of them. He knew the FBI was recording everything, and did not want to tell Mike T. because he wanted to hear how he whacked the kids. Fredric drank some more wine and listened as Mike T. continued his story.

"So the next day rolls around, which is Friday. I have some of my goons out there scoping out these kids and following them around. Their buddy Carmine picks them up and brings them to my house. It's amazing what people will do for money. He tells them that they're going to a basketball game and they have to pick up one other person. This other person happens to be me. So they come into the house and into my office where I'm sitting at my desk like Brando in *The Godfather*. When they see this, they realize I'm the guy they have to pay. I love watching people pee their pants. This Belkoy kid has a stream running down his leg and it's leaking through his jeans. I throw some tissues at him and tell him he better not ruin my three-thousand-dollar throw rug from the Orient, or I'm gonna whack him on the spot. I tell Carmine to leave so me and the two kids can talk business. As he walks out, my two hooligans step in. I tell these kids that we have a small problem. They owe me fifteen grand and they ain't got it. So I ask them what they think I should do. This Mindrago kid tells me we could just let it go and they would do community service for me, if you know what I mean. And I'm thinking to myself, do I really want two kids who bet fifteen grand on Tiger Woods to work for me? Hell, no! So now I tell them that they'll have the money for me by tomorrow at five. I'll meet them down at Penn's Landing and they'll give me the cash. Well, these kids are kissing my ass, telling me how nice I am and how forgiving I am, and I say, well, don't start buttering me up just yet. I don't *know* if you'll have it or not. So I need some collateral. I think a finger from each of you will work. Then I'll store it in ice and if you bring the money tomorrow, you'll get your fingers back. So I'm gonna go get a drink and when I come back, you'll tell me which finger you would like to lose, and I'll give you the butcher knife and you can do it yourself. I get a little shaky around those things and I wouldn't want to chop off their hand.

"You should have seen their faces! Their jaws hit the floor and the kid who peed his pants starts crying. So I go out and come back with a bottle of Wild Turkey, a few glasses, a chopping block, butcher knife, a bucket of ice, a black Sharpie, and a roll of paper towels."

Tony jumped into the conversation as he was cleaning out his gun with the table linen. "So what do you need the marker for, Mike T.?"

"It adds to the effect. I told them that all fingers look the same once they get chopped off, so if they wanted their own finger back, they would have to write their initials on it. So I come back with all the stuff and I pour each of them a glass of Wild Turkey. They slam it down like apple juice. Then they're sitting there trying to decide who's going first. They were shaking like jive turkeys at a Black Panther meeting. The Mindrago kid goes first. He chose his pinky finger. It was actually a pretty clean cut. Most people do a half-ass job and the knife gets stuck in the bone; very painful and a lot of screaming. He yells out a few obscenities and wraps his finger up. As he's sitting there crying, the Belkoy kid steps up to the cutting board. He's already crying and sobbing. I smack him in the face and tell him to be a man. The knife comes up and he swings with force. He misses his finger. Second time is the same thing. Finally, Anthony grabs the knife and does him a favor. He screams bloody murder and wraps his hand up and falls to the ground. I tell him to stop the drama and he better keep his hand above his heart or he's gonna bleed to death. Of course, that really scares him and he starts hyperventilating in between his cries. Well, I couldn't help but laugh, and so did the guys. I can't help it. It was funny. So with that I sent them on their way. Now I have to ask you guys, do you think they paid up the money?"

Jimmy, who was polishing his barrel, was the first to respond. "Of course they didn't pay, or you wouldn't be telling us this story. You know, we never tell stories about the people who paid. It seems like we're always talking about the people we kill. They do make for better stories. So Mike T., have I answered your question correctly?"

"Yep, you have, Jimmy. You're right. Why do we always talk about the guys we've killed?"

"Because it's like a bragging contest," said Tony. "We like to show each other up and stuff like that."

"It is like that, isn't it?" nodded Mike T.

"So what happened to the kids?" asked Fredric.

"Well, you know, they didn't have the money. They didn't show up on Saturday and I had to send my guys out to pick them up. They were hiding out at the Turf Club downtown. They were trying to make their money and were losing terribly. Once they saw my guys, they just walked over to them. Christ, if that were me, I would have hightailed it out of there. Belkoy started crying as soon as they got in the car. Of course, he soiled himself, and the car had to be cleaned to get the stench out of there.

"So, they show up at my house and I meet them outside because I didn't want my house stinking like poop. They're standing there looking at me like I should be doing something. So I say, you guys like hiking? They look at me kind of confused and everything. I ask them again if they like hiking. They nod. Then I tell them that if they do this thing for me, they can walk away scot-free. Their eyes light up and they're thinking I might actually let them walk away with their lives. I love lying to gullible people. I give them a briefcase and tell them it's full of money and they have to drop it off in the middle of Fairmount Park. I threw some old papers and stuff in the briefcase so it felt like something was actually in there. I give them two flashlights and my guys dropped them off along the Wissahickon.

"After they leave, I drove down there. I told the guys to take the long way so I could get out there before them. I know that place like the back of my hand. I've buried a lot of bodies out there. So I'm lying in wait as I see the flashlights coming down the trail. I can hear them talking to each other. I don't think they could've been any louder. The Belkoy kid was trying to convince the other guy to bust open the briefcase, take the money, and skip town. The other kid starts yelling at him and calling him names. Apparently, Belkoy was the one who made the bet and got them into trouble.

"So as they get to this spot, I come out of the woods and shine a flashlight under my face to try to look spooky. I scared the shit out of them and they just stood there screaming. I tell them to shut up and put the briefcase down. I think the Mindrago kid realized first that I was going to kill them. I could see his face turn pale as I shone the light in his eyes. The other kid started crying and hyperventilating again. I was sick and tired of hearing that wuss crying. He went first. I kicked him in the chest and he rolled on the ground. I took out my gun with the silencer, you know the one DeLuca gave me for my birthday last year. I shot him in the back of the head and that was the end of him. I turned to Mindrago and he had these pleading eyes, like don't kill me. I told him it's nothing personal. So I turned him around and shot him in the back.

"Now comes the fun part. I pick up my shell casings, put on some gloves, and position the bodies facing each other. I take out a note and place it in Belkoy's pocket. It says that he was in love with Mindrago's mother and the only way he could be happy was to kill his friend and run away to the Bahamas with the mom to live in sin. Then I wrote some stupid stuff, like tell my mommy I love her and blah blah blah. Then I take out another gun and stick it in Belkoy's hand to make it look like he shot himself after he shot Mindrago. I figured if this kid cried at the drop of a hat, then he wouldn't be able to live with himself if he killed his buddy. I was right, because the case never made it beyond the homicide unit. I typed up the paper on the computer, so there was no handwriting to compare. I was out fifteen grand, but, hey, I got two more fingers to hang in my collection."

Thy guys hooted with laughter and shook their heads. "You're one crazy mofo," joked Tony. "I'm glad I don't owe you money."

"It's just fun," laughed Mike T. "So, Fredric, why don't you tell us about one of your hits?"

Fredric stiffened. In his subconscious dream state, he had plenty of stories to tell about guys he whacked. He also remembered that the FBI was taping everything going on there. He didn't want to be implicated, so he excused himself to go to the bathroom and promised a story when he returned. As he got up, Tony stopped him. "Hey, don't forget your gun, There are some crazy French people out there."

They all laughed and Fredric did, too. As he went into the bathroom, an FBI agent came out from the stall. Fredric looked in and saw that the stall was set up like a private listening room. There were tape recorders, and a couple of guys sitting there with headphones on, smoking cigarettes, and drinking coffee from Styrofoam cups. Fredric was amazed that they could fit all of the people and equipment in there. He turned around to find Mike T. standing there with his gun out. "I knew you were working for these guys. How could you turn on us? You were the best man in my wedding. I hate to do this, but you gotta go, my friend." He raised his gun...

Fredric jumped from his desk. He looked around wearily, and as his head cleared, he realized he'd been dreaming. The chief was standing next to him. "Time to wake up, sleeping beauty. You've been out for two hours. I figured you needed the sleep, so I didn't wake you. Don't worry, kid, it might be the first time you fall asleep working on a case, but it definitely won't be the last. Welcome to the club."

"Oh no, Chief, I wasn't sleeping, I was testing my keyboard for drool resistance," he said as the chief went back into his office. Fredric rubbed his eyes and looked down at his desk. As he recalled details of the dream a light bulb went on in his head. His thoughts started coming together and he remembered Mike T. and an undercover investigation with the FBI. He read through the rest of the report on Mike Tamasco and learned that the FBI had been tapping his phone conversations. He smiled with excitement; he now had a new lead and a possible perp to place behind bars.

"Hey, Hassie, it's around five-thirty. Why don't you go home and get some sleep," offered the chief.

"Wait, Chief Cahill, we have a new lead. I think I know who killed those two kids over in Roxborough. Have you ever heard of a guy by the name of Mike Tamasco?"

"The name actually does ring a bell. I remember some of the guys down at the south precinct mentioning his name. Yeah, he has ties to the mob and DeLuca. They think he's responsible for several hits, but never had substantial evidence to bust him. What do you have?"

"Well, Chief, I was looking through this notebook that documents all of this kid's bets. As I was flipping through, I found this scribbled in the margin in the back." Fredric showed him the writing. "Now look at these bets, pretty much all of them say Gizep. That's the kid we busted at the poker house the other night who just walked out of here. He was their bookie and they made their bets through him. If you look toward the end, there are all of these entries for college basketball games. But the final championship game has no name, and this Belkoy kid lost his shirt. There are dollar signs with every entry to show how much he won or lost. If you look at the entry for the final game, which is also his last entry that coincides with the approximate date of the disappearance of Phillipe Mindrago, there are fifteen dollar signs there. That's fifteen grand! There is no name next to the entry, but then I find this Mike Tamasco name. What do you think of all this?"

"Good work, Hassie. It sounds like you have a new suspect. However, you have an FBI investigation going on. Those guys don't like anyone interfering with their business. We can get together with them and find out what's going on. They might let you in if you have some information to share. I would say go over to Monsternick's house and get a plan together. Let me know what you come up with. Actually, I guess I can tell you that I do know the FBI is on the case. They are doing some recon work right now with phone taps. Now, Hassie, I don't know if you know this, but acquiring a wiretap in Pennsylvania is about as hard as finding a sober driver on the road after two in the morning on New Year's. It takes all kinds of paperwork that has to be approved. It's hard for the FBI, so you can imagine how hard it is for us to get warrants for that stuff. But they are making solid progress and when the time is right, I will fill you in on the situation. Until then, keep your mouth shut, go rouse your partner, and see what you can come up with. If it's good enough, I might be able to get you in with the feds. It'll be an experience you'll never forget."

"Thanks, Chief," replied Fredric. "We'll keep you posted." Fredric's stomach rumbled with hunger. He pulled up a memory from a few months back when Melinda had shown up at his door early in the morning with breakfast and coffee. Now it was Fredric's chance to return the favor. If he left now, he could be at her house by six and hopefully avoid the morning deluge of maniacs on the expressway.

ᔕ   ᔐ

Melinda answered the doorbell still in her pajamas. She had one eye open and her bed head was screaming for a brush. Even though she had just woken up, she still had a cute face that any man would want to know. She had dated, but could never find the right man. They were either too self-involved or they were too insecure. Plus, Melinda could not find a guy who could hold an intellectual conversation for longer than 10 minutes. After that, it was just cheesy lines, or excerpts that bragged about his accomplishments. Instead, she had her cat Will, who weighed close to 25 pounds. She took care of him and he took care of her. It was a beautiful platonic relationship that only a cat lover could appreciate. All others would scoff and poke fun because they owned dogs.

"Hassie, it's six in the morning! What the hell are you doing here?"

"Rise and shine, Monsternick! I brought breakfast, and some terrific news. Carmine or Jo Jo didn't kill those boys. It was a guy named Mike Tamasco. He's a mobster from the city and the FBI is tracking him right now. We need to put a plan together so we can work with them and bust this guy. You should see his rap sheet, it's twenty pages long. This guy has been going to jail since he was a teen. He..."

"Stop!" said a flustered Melinda. "Come inside and let me put some clothes on. Have a seat at the kitchen table and give me a few minutes to wake up."

Fredric sat at the table and could not control his exhilaration. As he sat sipping his coffee, his mind was running full speed. His power nap at the station had revved him up and the java pumping through his veins only made it worse. As he looked around the kitchen, he noticed that Melinda's house was in immaculate condition. Everything was so neat and organized. Fredric wished he could be the same way. He was a little reluctant to pick up a mop and dustpan. Being a bachelor was fine for him; only if he had to impress someone would his sense of domesticity kick into high gear.

Melinda walked into the kitchen wearing a pair of shorts and a t-shirt that read, "The proctologist called...they found your head." Fredric laughed. "Where did you get that shirt?"

"My brother got it for me. He gets these crazy catalogs with random shirts, hats, and stuff. He bought it for me because my dad always used to tell him to get his head out of his ass. It's more of a personal joke than anything, so I wear it around the house. However, it's probably the common mantra of the general public."

"That's great. I love it," said Fredric. "So, pull up a chair and let me tell you all about my evening."

Melinda fixed her coffee and helped herself to a bagel as Fredric outlined his new hypothesis. "Well, let me start with my dream. I was sitting in Le Bec Fin with a mobster named Mike T. and his cronies. We were sharing stories about whacking people and we all had our guns sitting on the table."

"Wait," said Melinda. "When you say Mike T., do you mean the mobster-slash-hit man, Mike Tamasco?"

"Yeah, how did you know that?"

"His name has come up in several conversations with the detectives from the other precincts. Where did you come up with his name?"

"Believe it or not, it was in the Drater Club notebook. I was looking through it last night, and as I was flipping through the blank pages, I found one page that had a little scribble in the margin that said, 'Find Mike Tamasco.' I punched it into the computer and the printer spit out a small novel. This guy started off stealing cars when he was a kid and then he went to live with—get this—Johnny Merlotti."

"Yeah, I know about him," said Monsternick. "He's his uncle, right?"

"You knew that, Monsternick? Why didn't you tell me?"

"Because he was never part of any of our investigations together."

"Oh. So this guy must be one tricky dick because the FBI has a tap on him."

"Do you have the report?"

"It's in the car. I fell asleep while I was reading it last night. Let me tell you about this weird dream. Like I said, I was sitting in Le Bec Fin with them at some private table. I've never been inside there, but we had our own little room and we were sharing stories about guys we killed. Somehow, I was one of them and I had whacked some people, too. So we're sitting there swapping stories, and this Mike Tamasco guy starts talking about Belkoy and Mindrago. He tells us that they bet all this money on Tiger Woods in the PGA Championship. They lose and it turns out that Carmine DiGuiseppe put the bet through Mike because he couldn't cover it. So now the boys owe Mike fifteen grand and they can't pay it. He brings them to his house and makes them each cut off a finger!"

By this point, Melinda raised an eyebrow as the coffee made its way through her blood. "This is really random, Hassie."

"You think that's bad, the other guy was beaten with a hockey stick and ended up buried in ice with a hockey puck stuck in his mouth. So, back to my story. The kids never come up with the money, and he sends them out to deliver a briefcase in the middle of the woods. He goes out there and scares them to death, then shoots them and leaves some silly note about Belkoy falling in love with Mindrago's mother and wanting to move to the Bahamas."

"You were pretty tired, weren't you?" asked Melinda.

"Exhausted," he replied.

"It's not the last time that will happen. I've done it before and conjured up some pretty weird stuff. So, Freud, how do you interpret your dream?"

"Well, here's the strange thing. I knew the FBI was tapping our conversation, so I walk into the bathroom, and the feds are standing in the stall running their taps. I turned around and Mike raised his gun, and I woke up. Monsternick, we have to figure out a way we can infiltrate ourselves into their investigation. They are tapping him to find information about Frank DeLuca and his supposed construction company. I don't think they know that he could possibly be the killer of the two boys. If we can find something, we might be able to work with them. What do you think?"

"The feds are tough. I'll tell you what. We had a case one time where we did this huge drug bust. Some dude had, like, a million dollars worth of cocaine that was shipped up here in a car. The car was filled with cocaine everywhere imaginable. The tires had cocaine, the doors, the ball in the stick shift, even under the dashboard. You had to do all this crazy stuff to get it to open. Anyway, we made the bust and called in the feds for back-up. These guys come out of the woodwork. I mean, they were everywhere. I don't know exactly how many agents we have in Philly, but there are a lot; like twice the size of the Philadelphia Police Department. Hassie, it's going to be tough getting in with them. If they have claimed their jurisdiction on this guy, we need a mighty good reason to be a part of their team. I think it's a male-ego thing where they think they're better than us. They receive all this fancy training, and that's supposed to make them supercops. Well, I've met a few who would have their nuts bronzed if they could." She pointed to her shirt and smiled. "A lot of these guys out there."

"Before we go messing with the feds, let's do some homework, Melinda said. "I have some old informants in south Philly, and I am sure they'd be able to give us some kind of scoop. If we can find out some good dirt, we might be able to go to the feds with it and possibly set up a pickup and some questioning. The only thing is that they might want to make the collar and take the credit. They might also want to do the interrogation, in which case, they will probably angle it so they end up asking questions about Mike T's ties with DeLuca and we won't have anything about Belkoy and Mindrago. Our third choice is that they decide it's too risky and they won't let us do anything. We could be shut out from the investigation. We need to find out what Mike T. does as a cover for business. He has to do something during the day. It's pretty difficult to kill someone when the sun is out. Did you talk to the chief?"

"I told him about it and he seems to know more than we do. He was not at liberty to discuss the case. He did mention that if we can come up with something solid, we stand a good chance of working with the feds. I'm so pumped about this case, Monsternick. I want to go out right now and find your informants. Then we can set something up on Mike Tamasco and nail him."

"Not so fast, my friend," cautioned Melinda. "First things first. I need to shower and I have to be in court for the next two days. Friday we can take a ride down to the docks and see what we come up with. Good thing it's payday this week, Hassie. We might need a few bucks for information, if you know what I mean. They don't take checks either, so bring some cash. It's not the most orthodox approach, but it works. These guys are all recovering addicts of some sort and most of them haven't stayed on the wagon too long. If we can find one who is particularly hard up, we'll get him singing like a bluebird on a sultry day looking for love."

Fredric agreed and left Monsternick to tend to her duties with the justice system. He decided to go home and do some online research.

Seated in front of his computer, Fredric punched in the name Mike Tamasco and many articles surfaced about his involvement with the mob, past prosecutions, and courtroom fiascos. This guy had a pretty polished team of lawyers and possibly some politicians in his back pocket. In some cases, evidence was lost, misplaced, and even mishandled at the scene. Then there were investigations from the Internal Affairs department, because they thought the cops were on their payroll and compromised the evidence on purpose. All detectives go through procedural classes that educate them on handling evidence and maintaining its integrity. Some of these mistakes were blatant, like not wearing gloves while collecting DNA evidence, and the profile showed up with three types of DNA, one of which happened to belong to the guy who collected the evidence. It was apparent that Mike Tamasco had a lot of friends in the city of Philadelphia, and if he were to appear in court again, there would need to be an overwhelming amount of irrefutable evidence or some sort of smoking gun to put him behind bars. Fredric was determined to be the guy who was responsible for doing just that.

13

Winos and Wine Tastings

Friday morning found Melinda and Fredric riding to South Philly in a Mitsubishi 3000 that had been recently confiscated from a drug dealer. Detectives had the luxury of recycling some of the cars that were donated by the city's finest. All of the necessary paperwork was taken care of and the car now belonged to the taxpayers. The windows were tinted and the chrome wheels spun in the opposite direction of the tire. The inside was lined with black light tubing and leather. The stereo was hooked up to Sirius satellite, so more than 200 channels were available. Melinda preferred jazz, while Fredric wanted blues. They settled on a half-and-half schedule—blues to their destination and jazz on the way back. In the middle console was a TV screen that would spit out directions to anyplace in the world and in 25 different languages, to boot. If that was not enough, one could play Xbox or watch a movie with the DVD player. The car did have a nitrous tank hooked up in the back, but it was taken out and replaced with shotguns and crime scene kits. The ride was quite luxurious and the two of them enjoyed the new wheels.

As they turned onto Snyder Avenue, they saw that the Melrose Diner was bustling with all sorts of visitors. It was one of the more famous diners in the city and had quite an eclectic clientele. Fredric was keeping an eye out for a parking spot as Melinda watched some teen boys who probably should have been in school. They were pointing to the car and talking to each other. They either recognized it as the car of the now-captured drug dealer, or they were planning on stealing the shiny spinning mags. The boys sat watching as Melinda and Fredric emerged from the car dressed in shorts, flip-flops, and t-shirts. The boys snickered at them from across the street, and Melinda could hear their conversation.

"Yo, she driving Big Mo's car. That ain't right, man. He done got busted two weeks ago and they driving his car already. Yo!" he yelled from across the street. "I know that ain't your car. It belongs to Mo, and once he get out, he gonna want it back."

Melinda cracked a cute crooked smile. "Well, I don't think you're going to see Mo for a while. He's in jail for so long, this car will have antique license plates on it when he gets out."

"Oooh!" joked the boys and started pushing each other around. "She know all about Mo. You two must be five-O, right?"

"Who, us?" joked Fredric. "Nah, we just bought this at an auction. Who's Mo?"

"You can't jive us, bro. If you were any more popo, you'd have a camera crew filming an episode for *Cops*."

"Well," said Melinda, "I guess you're just too smart for us. Say, shouldn't you be in school right now? There's one more week left before they let out. How about we come over there and check some IDs?"

"Hell, no!" yelled one of the boys and they dispersed from the corner.

Melinda and Fredric focused their attention on the diner. They were not there to have breakfast, but to look for a guy who called himself Lightning Hopkins. No one knew why, but he had been in and out of the system, including rehab. Melinda had been using him for 10 years now and he'd proven to be one of her best informants. He was a scraggly man in his early 40s who looked much older from his years of drug and alcohol abuse. Sober moments were seldom, but for some reason, he remembered more when he had a buzz going. They turned the corner to find Lightning raiding the dumpster behind the diner. There were some fantastic scraps in there, and if you hit the dumpster at the right time, you could have Eggs

Benedict with toast and sausage. For a Friday morning at 10 A.M., the streets were not as busy as usual, probably because it was the Friday before Memorial Day weekend. Melinda put her fingers in her mouth and let out a loud whistle. Lightning looked up with half of a bagel in his hand. He smiled, revealing the six teeth that were left in his mouth, and shuffled over to Melinda. "Hey, Monsthernick, what bringth you down here?" His lisp was a speech pathologist's nightmare.

"Oh, not too much. We thought we might join you for breakfast. This is my partner Fredric, but you can call him Hassie."

"Hathie, how are ya? Nithe to meet you. My name ith Lightning Hopkinth."

"That's an interesting name," replied Fredric. "Were you a famous blues player at one time?"

"Naw, I jutht got hit in the head with lightning one night down on the dockth. I had too much to drink and pasthed out with my metal pipe in my hand. It wath the damndesth thing. That little peathe pipe brought down the electithity like the almighty himthelf and gave me a jolt. Good thing I had my rubber tholes on, because it thaved my life. If not, I would've been a goner. Ever thince, they been calling me Lightning Hopkinth."

"That's one hell of a story," said Melinda. "Say, buddy, I got some friends here who are looking for a little information." She waved a few $20-dollar bills in the air.

"Monthernick, thep into my offithe behind the dumpthter."

Melinda knew that he had a reputation to protect and the last thing he needed was for someone to see him with them and get killed for being a stool pigeon. Melinda made sure to canvass the area before approaching him. She had bypassed him on previous days because there were too many suspicious people around. Today the streets were quieter than usual.

"Tho, who you looking for?" Lightning asked.

"Have you ever heard of a guy named Mike Tamasco? I think he's one of the locals down here." Melinda waited as Lightning scratched his stubbly chin with his dirty fingernails.

"Yeah, I know that guy. He run a plathe down on Packer. I think it number 583. He do food and sthuff. I know he doing grapesth from down in Thouth America. I go over there for free wine. He got it thet up at a table in a big keg. He a big man around here. Everybody know Mike. He give out free turkeysth at Thanksthgiving, too. He give a lot back to the community."

"Do you know if he runs bets for anyone around here?" asked Melinda.

"He thure do. I ain't never bet through him, though. He kick you ath if you don't pay. Word ith that he taken out more than a few throughout the thity. He never get caught, though. He got thome good lawyerth. I ain't gonna talk about him no more cauthe he one bad dude. You don't meth with him."

"Thanks, Lightning," nodded Melinda with a smile. "We won't bother you any more. Here's a few extra bucks. Go get cleaned up and have a good meal on us. Don't spend it all in one place, if you know what I mean. Take care, buddy, and try to stay sober."

"Yeah, Monthernick, like that might happen." He let out a gritty laugh that lead to fit of coughing that sounded like a smoker on his deathbed.

Melinda and Fredric made their way back to the car. Fredric was laughing to himself. "Monsternick, I'm surprised that guy didn't die on us back there. How does he survive?"

"You'd be surprised how many like him are roaming the city. We have a serious problem with homeless people. You really don't think about it until you start walking the beat."

"I know, I just came from street duty, remember? I mean, that guy was in pretty bad shape."

"I've known him for ten years now. He's never steered me wrong. I busted him several times when I first started working for the department. He was holding up convenience stores for drug money. Now he gets skimpy welfare checks that he drinks or shoots up his arm. It's a damn shame, but we'll never be able to change him. Instead, we throw him a few bucks in exchange for some information and hope he spends some of it on something useful." Melinda sighed. "It looks like we'll have to take a trip over to 583 Packer Avenue. Are you in the mood to drink some wine?"

"It's a little early for that," returned Fredric. "However, I do remember my days in college when we would wake up at eight in the morning and start tailgating for football games. God bless the University of Nebraska and the sea of red." Fredric closed his eyes for a second and reveled in the glory of his days as a carefree college kid.

"Snap out of it, Hassie," Melinda said as she smacked him on the back. "Let's go check it out."

∾ ∾

The warehouse was situated at the end of the street. On the corner was a big sign that read "Tamasco's Produce, Inc." The parking lot was full of Italian men in construction trucks loading up crates of grapes. Lightning had not been kidding about the new shipment. There were several forklifts operating and bringing out tons of grapes to be loaded into the beds of the trucks. Fredric laughed because he thought most construction guys were beer drinkers. "I thought these tough guys would be buying malt, hops, and barley for some brew. Who knew winemaking was such a blue-collar hobby?"

"Surprisingly enough," interjected Melinda, "there are a lot of guys who make their own wine down here, especially in South Philly. When I worked down here, some of the guys on the force had traditions of winemaking passed down from generation to generation. I tried some of it, and it was excellent. These guys on the force had a little wine club called California Dreamin'. They had all of their grapes shipped in from Sonoma County in California. If you know anything about wine, which I know you don't, it's one of America's greatest growing regions for grapes. Their wine tasted like it was right from the store and it was strong. The best part was that you never had a hangover because they never pitched in any sulfites. That stuff will give you one nasty headache."

"Okay, enough with the wine jargon. I have no idea what you're talking about. Maybe one day when I get the acquired taste. Until then, I'll stick with beer. So, do you think a lot of these guys are part of Mike Tamasco's little posse?"

"It's hard to say," replied Melinda. "I don't recognize any of these guys as major mobsters. They're just some Italian construction workers picking up grapes. Let's go inside and see what's happening."

The atmosphere inside the warehouse was chaotic. Men in their construction boots were tagging crates of grapes for purchase. Others were standing around tasting grapes and pressing their juices on a little instrument called a refractometer. It gave a reading of how much sugar was in the grape. Apparently, it was important to the alcohol output. The warehouse was huge and every corner had something different. Aside from the grapes, there were all sorts of winemaking equipment, from grape presses to oak barrels and glass containers. In another corner was a long table set up with several small wooden barrels. Each one was labeled with a different wine from wineries in the area. There were also bottles on the table that people would drop off for others to try. Of course, this area was crowded as well because the wine was free and accompanied by various cheeses, crackers, and bread. There seemed to be more women in this area. Melinda led the way and Fredric followed as she picked up two plastic cups and started to pour. "Fredric, let me give you a short lesson in wine. These barrels here are used to add oak flavor to the wine. It makes all kinds of changes from the taste, to the way it coats your mouth, to the taste after you swallow and exhale."

"Monsternick, I really don't care. I don't like this stuff. It's pulling on my tongue like a sour candy."

"That's because it's been in there too long."

"Yeah, I agree," commented a woman who was also puckering up her lips. "I need some cheese to cleanse my palette."

"Me, too," agreed Melinda. "This place is a serious production."

"Oh, yes, Mike runs quite a business for winemakers," the woman replied.

"Mike who?" questioned Melinda.

"Why, Mike T. Everyone knows he gets the best grapes in the tri-state area. We never buy from anyone else. In fact, we belong to his club, Vendemmia Di Tamasco."

"That sounds like an interesting club. How does one become a member?"

"You have to be nominated by an existing member. My husband is a lawyer and has represented Mike T. in the past. He's one hell of a guy."

"Mike T. or your husband?" asked Fredric.

The woman laughed and poured another glass. "Mike T., of course. Don't get me wrong, I love my husband, too, and he is a great man, but Mike T. has done so much for the club and even the community. He has donated money to so many organizations and helped the kids build a new playground. He is a true saint."

"He sounds like a fantastic guy," commented Melinda. "Is he here today?"

"No. He comes in and out and I think he's out of town for the morning. His partner Lenny is running the show. He's over there behind the register."

They looked over to see a middle-aged man in a sweaty baseball cap with a cigarette hanging out of his mouth. It was obvious that he was helping himself to the wine as well. He had a glass next to him and he hoisted it up for a gulp.

Melinda returned her attention to the half-lit lady stuffing her mouth with shreds of Romano cheese. "So this club, do you get together at Mike's house and have parties?"

"Sometimes," she said. "Actually, tonight there's a huge celebration over at Davios Italian Steakhouse on Broad. He rented out the whole place for our club. We're celebrating the first bottling of the year. We get together every year at this time to share our wines. The spread is amazing and the wines are fabulous. My husband usually places in the top three. He has some pros working with him who have access to all kinds of equipment. Too bad you guys aren't in the club. You could come tonight."

"Yeah," replied Melinda. She could detect the snobbery from a mile away. "It sounds like such an elite society that can only be enjoyed by people who really appreciate the finer things in life."

"Absolutely," she said. Then she looked both of them up and down trying to gauge their financial status based on their clothes. Deciding they were not of worth, she turned to pour another glass and went to schmooze with the other girls having a liquid brunch.

Melinda and Fredric left and sat in the car for a few minutes while the air conditioning cooled the car off from the increasing humidity outside. Melinda also changed the radio station and Fredric frowned. "I wonder if that lady knows who Miles Davis is?" Melinda asked.

"Do you think her husband is the guy I've been reading about in the report? He sounds like a real jerk."

"Well, if he married that cheesy beach ball, then I would venture a guess in the affirmative."

Fredric cracked a smile. "How weird is it that we found out about Mike T. without asking anyone? That fell into our laps like a present from Bacchus. He's the god of wine, isn't he?"

"Hassie. I'm impressed. You do know something about wine. I'll convert you yet. Yeah, that was a huge bonus. I would definitely say something like that will probably never happen again. It's a one-timer."

"Let's head back to the station and talk to the chief. We should tell him what's going on and see if we can get a shot at picking Mike T. up tonight. If, and I say if, our luck holds, I might go out and buy a lottery ticket. This could be the break we're looking for. We have to be very convincing about our plan and make sure that Chief is on board one-hundred percent. He'll find a way to get us in with the feds."

❧   ❧

Back at the station, Melinda and Fredric updated the chief on their progress. He was very excited to hear about their escapade and the information about the wine dinner. There were some phone calls that had to be made, but other than that, the chief was pretty confident about getting the feds to let Melinda and Fredric have a crack at Mike T.

Sure enough, an hour later, the chief came back with good news. "I talked to the feds. They have a few stipulations. You can't ask Tamasco anything about his connections with the mob. They don't want to scare him off. Then he'll be ripping through everything looking for those wiretaps, and months of work will be blown. You can say that you know he associates with people in the mob, but you have no evidence on that.

Focus your questions on Belkoy and Mindrago and spin it so you think he's running his own gig. I don't want to hear the name DeLuca mentioned at all."

"You got it, Chief!" exclaimed Melinda as she jumped out of her seat and shook hands with him.

That night landed them on Broad Street standing outside of Davios Italian Steakhouse. "Monsternick, this food smells incredible. I didn't eat much dinner and I'm starving."

"Here, have a piece of gum. When we're done, I'll take you to Pat's Steaks."

Both of them were dressed up for their fancy evening at the wine celebration. As they walked inside, they were greeted by a waiter. "Good evening, may I help you?"

"We're here to celebrate the first bottling of the year for Vendemmia Di Tamasco," said Fredric in a confident and authoritative voice.

"Of course, sir. May I have your last name?"

"It's Petrozzo, Stephen Petrozzo. I am Mr. Tamasco's personal lawyer."

The waiter did not realize that the real Mr. Petrozzo was already there with his wife. He had no clue about the Mafia, and watching the news was not part of his daily routine. He quickly compensated for his ignorance." Of course, sir, I should have recognized you. Please accept my most sincere apology. Please enjoy your evening."

As they walked away, Melinda gave him an elbow in the stomach. "You remembered his name from the reports you were reading."

"Yes, I did. See? I'm learning how to do this detective stuff. You have taught me well, Miss Mayagi. Maybe later you can teach me to paint the fence and do the famous karate stance?"

"You got it, grasshopper."

As they looked around the room, men in tuxedos and women in evening gowns were seated at the tables. If they didn't know any better, they would have thought it was a gala for the mayor or something. The wine was flowing and food was everywhere. Trays of appetizers passed by filled with bites of crab and little soufflé cups filled with caviar. It was sheer elegance. Melinda wished she could stay a while and enjoy the food and wine, but the sight of Mike T. brought her back to reality. She nudged Fredric, who was helping himself to the jumbo shrimp. "There he is. Why don't you have a few more bites and I'll stand here and watch for a few minutes. Try not to talk to anyone. If they ask, we're friends of the Petrozzos."

Mike T. was decked out in a black silk suit with a pink tie. He was a very handsome man, and Melinda momentarily entertained the idea of being his girlfriend. Here was an intellectual who was interested in nature, hiking, and outdoor sports. They could travel around the world and soak up different cultures. Maybe even start a family, own a house outside the city with a small farm in the back...

""Do you want any of this?" offered Fredric, breaking into her reverie.

She shook her head. "No, thanks. Look, he's going somewhere. Maybe we can approach him in the hallway without making a scene. Are you done?"

"Yes. I could sit here for another hour, but I guess duty calls. Let's get this bastard."

Mike T. entered the bathroom. There were some lounge chairs out-side the restrooms and as Melinda and Fredric approached, they noticed the annoying lady from the warehouse sitting in one talking to another woman. Melinda stopped Fredric for a moment and turned him around. "We have to time this so we can get to him without her noticing us. If she sees us, we're going to have some problems."

When Mike T. emerged from the bathroom, Mrs. Petrozzo grabbed him and made him sit down between her and her friend. She went on and on, complimenting him about everything from his generosity to his silky-smooth suit. Melinda turned around and mentioned to Fredric that it was almost as disgusting as the way Fredric had flirted with Kristen the recep-tionist at National Medical Labs. He laughed quietly and pinched her arm.

As they watched, they could see Mike T. really did not want to have anything to do with Mrs. Petrozzo. He was looking everywhere else and even checked his watch a few times. The obvious hints didn't register with her as she rambled on. Finally, Mike T. cut her off mid-sentence and excused himself. She thanked him for his time and asked if he could come over to have dinner with her and her husband some night. Melinda saw it as a blessing, because several people had walked by and looked at them like they should not be there. The guests all seemed to know one another, but they didn't know Fredric and Melinda. It was time to make a move. As Mike T. walked toward them, Melinda let out a lovely smile, and he smiled as he got closer. Before he could extend his hand and introduce himself, Melinda cut him off. "Mr. Tamasco, I'm Detective Melinda Monsternick and this is Fredric Hassloch. We're from the Philadelphia Police Department. I don't want to flash my badge and make a big stink. Is there somewhere we can talk?"

Mike T. was taken aback. The smile on his face disappeared as he gestured for them to follow him to the back of the restaurant to the manager's office. On the way, he stopped to greet people and exchange a few words. Melinda and Fredric hung back so as not to interfere. They made their way through the kitchen and several people stopped and watched as they entered the office and closed the door. "How can I help you officers?" Mike T. asked.

"We're not officers, we're detectives," corrected Melinda. "We'd like to ask you a few questions."

"Please feel free to ask me anything you want. I have nothing to hide, but if you could keep it brief, I do have a lot of guests to entertain tonight."

"Your name is Mike Tamasco, right?"

"Yes."

"You run a food produce warehouse down on Packer Avenue?"

"Yes."

"Do you know an Adrian Belkoy or Phillipe Mindrago?"

"Never heard of them. Who are they?"

"They're two kids who attended Roxborough High. We found both of them dead. One was in Fairmount Park and the other was killed in his house."

Mike T. shrugged. "So why do you need to talk to me?"

"Have you ever heard of a Drater Club, Mr. Tamasco?"

"No, I haven't. Are we almost finished here? Obviously, I don't have any information regarding your case." He glanced down at his leg for a second than back at Melinda. "I really don't know what you're talking about."

"Do you know a guy named Carmine DiGuiseppe?" asked Fredric.

His posture became more rigid and he looked up at the ceiling. "No, I don't. Should I?"

He glanced down at his right leg again and Melinda caught his eye. "What's the matter, Mr. Tamasco? Is there something down there that's bothering you? Do you mind if we frisk you?"

"Is that really necessary? You're just asking me some questions. There's nothing suspicious about my answers. I don't see why you need to frisk me. I really need to get back to my party. Now, if you'll excuse me..."

"I don't think so," said Melinda. "Before you walk out that door, think about this. We could've made a big scene escorting you out in the middle of your party and embarrassing you in front of your guests. Instead, we chose to speak to you in private. Now I'm going to ask you this one time. If you lie to me, you'll walk out of here in cuffs. Do you have a permit here for that gun strapped to your leg?"

"Of course I do. It's out in my car."

"May we see it?" asked Melinda.

"Well, actually, I took a limo here tonight. Why don't you stop by the house tomorrow and I'll gladly produce the permit."

"Sorry, Mr. Tamasco," replied Melinda. "You've just given us probable cause. Without that permit, you'll have to hand that gun over to us. Once you bring the permit down to the station, you can pick it up."

Mike T. bit his lip and then reluctantly rolled up his pant leg and unstrapped the gun. Melinda pulled out a pair of gloves from her Kate Spade handbag, surprising both Fredric and Mike T. "See, Hassie, you always carry a pair of gloves no matter where you go." She took the gun, placed it in a plastic bag, and put it in her purse.

"I'll be in later tonight to pick up that gun. Do you really want to get into this?" asked Mike T.

"Are you threatening me?" Melinda raised an eyebrow at him.

He looked at her with dead-set eyes. As she stared back at him, she could see in him the coldness of a killer. "No, that's a promise."

"I'll make a note of that," smiled Melinda. "I'm sure we'll meet again. You have a nice night, Mr. Tamasco."

She opened the door and they exited through the back. Mike T. hastily returned to the restaurant in search of his lawyer, Stephen Petrozzo. He pulled him aside for some privacy. "Steve, the police were just here and they took my gun. We need to get that gun back ASAP. As soon as dessert is served, we need to go back to the house and get the permit."

Petrozzo frowned. "They can't just waltz in here and take your gun. That's illegal search and seizure. Well, they can't do much because the lab is closed right now. We can leave after the party and have it in a few hours."

"Right," agreed Mike T.

Melinda and Fredric jumped in their car. Melinda placed her hand-bag by her side. "We have to get out to National Medial Labs right now. Go through my phone and find Dr. Rieders' number in my contacts. Tell him he has to meet us there right away."

"Are we going to have him test it tonight?"

"Damn straight we are. We really didn't have the right to take that gun. He never signed a search and seizure consent form. We're taking a huge chance here, Hassie, because his lawyer will try to have it thrown out because we didn't follow the protocol. However, if we can prove our prob-able cause to take the gun to prevent others from being hurt, we stand a chance. Our other problem is the feds. After we call Dr. Rieders, call the chief and tell him what's going on. If he calls the feds and talks to them, it might not be as bad."

"Was that gun a .45?" asked Fredric.

"It was," she returned. "That's why we need Dr. Rieders to test this tonight. Tamasco is going to be coming to the station soon to retrieve his gun. If we can get the tests done and return it before he gets there, we'll have some samples to compare from the crime scene."

Fredric called Dr. Rieders and he agreed to meet them at the lab to run the tests. It was an unorthodox for him to work in the middle of the night, but he liked Melinda and wanted to help her out.

Fredric called the chief, who was silent for a few seconds. He was not too happy with their choice. "How am I going to explain this to the feds?"

"Tell them that we're running tests on the gun right now. Dr. Rieders is meeting us at the lab. We'll have the gun back in two hours. If they get there before then, we're screwed. If not, we might have some excellent evidence to nail this guy. Tell the feds that if the evidence does match up, we won't make a move until they're ready. It's more fuel to add to the fire. If they can't get him for his racketeering, we still have two murder charges we can slap on him."

"You guys are pushing your luck," said the chief. "I hope for your sake and mine that you make it back in time. Make sure Dr. Rieders doesn't fume the gun. We don't want any residue on there. We want him to think it's just been sitting in the holding room and we haven't taken it to the lab."

"Yes, sir," replied Fredric as he hung up the phone.

"The chief isn't too happy with us, but if we can make it out to the lab and back in time, we're safe. Otherwise, we're up the creek without a paddle. Are you sure Dr. Rieders will be able to do this quickly?"

"Look Hassie, all we need are some test bullets, gunshot residue samples, and possible prints to run through AFIS."

Dr. Rieders was waiting for them at the lab. As they went back to the firing room, Dr. Rieders assured Melinda that he would be able to get all of the samples, but they would have to wait until tomorrow for the results. He also had a chance to use the alternate light to find and photograph some prints to run through AFIS. All of the instruments were shut down, and there was not enough time to start them up and wait for samples to be run. He promised results in the morning.

A half-hour later, the samples were collected and placed in the appropriate evidence bags. It was nearly midnight, and the gun needed to be back at the station soon. Dr. Rieders resolved to pull an all-nighter for them and fired up the ICPMS. "It's been a long time since I've stayed up all night. This reminds me of my days of youth studying with the medical examiner up in New York City. I loved the graveyard shift because that's when all the crazy stuff came in. If you've ever heard that saying, 'the freaks come out at night,' well, it's true. Earth is the insane asylum for the universe."

"That's great, Dr. Rieders, but we really have to get back to the station with this gun. When we solve this case, I promise we will take you and your lovely wife out to dinner and listen to your stories all night. I truly treasure them. But for now, we have to make like a banana and split back to the city. Thank you so much. We owe you one." Melinda gave him a big hug and slapped a kiss on his cheek. He smiled and nodded.

The two hopped back in the car and made record time heading to the station. Mike T. and his lawyer showed up five minutes later to claim his property. He inspected the gun and figured that it had been sitting there the entire time. He smiled, thinking that no samples could have been taken from his gun. Boy, was he wrong.

ᔛ   ᔐ

Vocabulary

Charge   A fundamental characteristic of matter, responsible for all electric and electromotive forces, expressed in two forms known as positive and negative.

Chromatography   A method of finding out which components a gaseous or liquid mixture contains, which involves passing the mixture through or over something that absorbs the different components at different rates.

Ions   An atom or group of atoms that has acquired an electric charge by losing or gaining one or more electrons.

Mass spectrometer   An instrument used to separate compounds and measure their molecular weight.

Mobile phase   The components of a sample become soluble and travel, leaving marks of separation.

Molecule   The smallest physical unit of a substance that can exist independently, consisting of one or more atoms held together by chemical forces.

Retention time   The time it takes for certain molecules to complete the flow through a column.

Stationary phase   The absorption of material into the chromatographic medium.

Toxicology   The study of the nature, effects, and detection of poisons and the treatment of poisons.

Background Information

Chromatography enlists the help of a forensic toxicologist. Toxicology is the study of the nature, effects, and detection of poisons and the treatment of poisons. In the crime lab, the toxicologist is concerned with identifying unknown substances, whether they are in bodily fluids, water, air, or everyday products that can harm an individual or the environment.

Chromatography is one of the tools used in the laboratory. It separates and identifies individual chemical compounds found in a solution, gas, or solid. Chromatography was discovered by a Russian scientist named Mikhail Tsvet. It is based on the theory that chemical substances have a tendency to partially escape into the surrounding environment when absorbed on a solid surface or dissolved in a liquid.

There are two stages that are important to chromatography. The first one is the stationary phase, in which the material absorbs the components of the mixture. There are several mediums that can be used for the material, such as paper; thin layer chromatography, which uses thin glass plates; gas; and liquid. All of these work on the basic premise of separating a component through absorption and separation of the compound. The other phase is the mobile phase, which is used to carry the components or cause them to move along the instrumentation. Time and time again, measurable results are achieved in the form of color chromatograms or a computerized chromatogram that shows a series of peaks and valleys.

For example, if a chemical is dissolved into a liquid, some of the molecules will travel faster than others. Eventually, they will totally separate from each other and the chromatographic process will be complete.

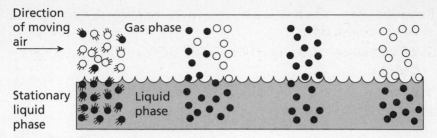

Chromatography in gas and liquid phases.

It is similar to a race between the chemical compounds. As the race begins, those molecules that have a higher affinity for the mobile phase slowly pull ahead of the others. At the end of the race, all of the separated compounds cross the finish line at different times, labeling them as being derivative of specific compounds or molecules.

Gas chromatography (GC) and high performance liquid chromatography (HPLC) are commonly used in the laboratory. They are favorable because they can separate a mixture into its components within minutes. With GC, the mobile phase is a carrier gas. This gas flows through a column constructed of stainless steel or glass. The stationary phase is a thin film of liquid contained within the column. The columns are anywhere from 6 to 60 meters in length and 3 to .25 millimeters in diameter. As the carrier gas flows through the column, it carries along the components that have been injected into the column. The components with a higher affinity for the gas travel faster through the columns.

The path is as follows: First, the carrier gas, which is usually helium or nitrogen, is fed into the column at a constant rate. The sample is injected as a liquid into a heated injection port with a syringe. Upon injection, it is vaporized and swept into the column by the carrier gas. The column itself is heated in order to keep the sample in a vapor state. As it travels through the column, the components of the sample separate. Upon finishing the process, the detector records the time it takes for the samples to finish and creates a chromatogram. The time it takes for the components to emerge from the column is called retention time.

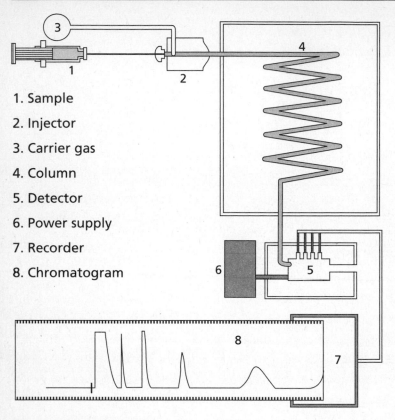

1. Sample

2. Injector

3. Carrier gas

4. Column

5. Detector

6. Power supply

7. Recorder

8. Chromatogram

The chromatographic process.

Chromatography is often coupled with mass spectrometry. Chromatography is excellent at separating compounds, but one major drawback is that it does not identify the substance as being one compound or another. A mass spectrometer measures the mass-to-charge ratio of individual molecules that have been converted into ions. Once that takes place, the information is used to determine the masses of the molecules. All elements have specific molecules that will register on the spectrometer with their respective mass. As compounds exit the gas chromatograph, it enters a high-vacuum chamber where a beam of high-energy electrons collides with the molecules and causes them to lose electrons and acquire a positive charge. These are also called ions. Ions are

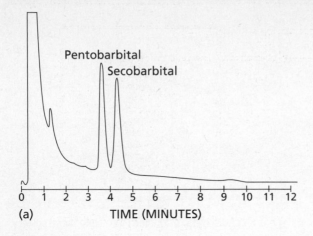

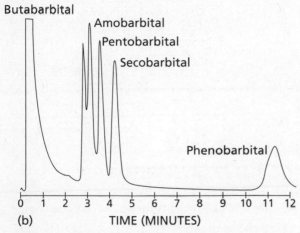

An unknown mixture of barbiturates (a) can be identified by comparing it to the retention time of a known sample of barbiturates (b).

very unstable and almost instantaneously decompose into numerous smaller fragments. These fragments pass through an electric field where they are separated according to their masses. After the fragments are recorded on a mass spectrum, the substance is easily identifiable, because it is measured with a database of other mass spectrums for chemicals and their elements. Heroin, for example, always separates the same way and has similar spectra compared to other samples. Once the sample is compared with the database, the compound is identified.

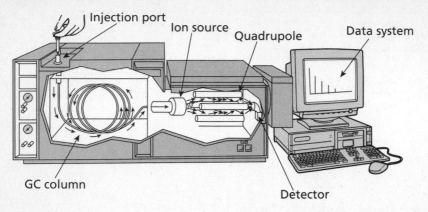

Diagram of inside elements of gas chromatography coupled with mass spectrometry.

Mass spectrometry has also proved useful in identifying gunshot residue. This has been employed with inductive coupled plasma mass spectrometry (ICPMS), which uses hot plasma to separate the sample before it goes into the mass spectrometer.

Forensic Implications and Jurisprudence

Gunshot residue (GSR) consists of different powders that ignite when charged by a spark. Some of the elements that may be found in GSR are copper, nickel, lead, barium, antimony, and tin. It starts with the primer that is used for initial fire. The primer exists in two types of fire. The first is rim fire, which means the primer is in the base of the cartridge. If any spot on the rim is struck by the firing pin or hammer, the cartridge ignites and propels the projectile forward. The other type of primer is called the center fire primer. In this instance, the primer is centered in the base of the cartridge. When the firing pin strikes the middle of the cartridge, it causes the projectile to fire.

The combustion of cartridge powder produces gases that propel the projectile from the cartridge and through the firearm. Pressure emanating from the expanding powder within the cartridge is sufficient enough to

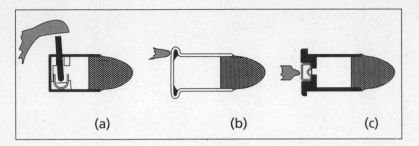

Pin fire cartridge (a), rim fire cartridge (b), center fire cartridge (c).

force the case against the breech face and the chamber walls. This is what makes the cartridge unique. It can then be matched to other cartridges from the firearm.

Black powder, composed of 75 percent potassium nitrate, 10 percent sulfur, and 15 percent charcoal, used to be common. Black powder proved to be inferior because when it combusted, it produced a great deal of black smoke and damaged the inside of the gun barrel. Smokeless powder, introduced around 1866, has become the standard in all ammunition.

Gunshot residue may land on the hands or other body parts, clothing, or any other object that is in close proximity to the weapon.

As discussed in Chapter 4, there are three types of gunshots: close range, distant, and direct contact. A shot is considered to be close range when residue is found around the wound or shows on a surface, like clothing, that the projectile penetrated before hitting its target. Close-range gunshots show a marked area of GSR. As the shooter increases the distance between himself and the victim, there is less GSR. The type of weapon and ammunition used also significantly change the distances GSR travels. Obviously distant shots have no visible GSR where they strike their target. A direct-contact wound has significant amounts of GSR in and around the wound.

At 2 to 4 inches, the exploding gases from the muzzle are hot enough to scorch and sear. If the scientist conducts a microscopic examination of the area around the entrance wound, skin may be burned, hair may be singed, and fibers of cloth show thermal changes.

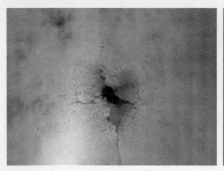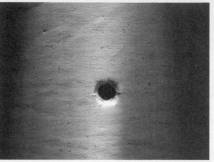

A 9mm bullet fired through cardboard at 6 inches (left) and at 12 inches (right).

As the distance increases, scorching is not apparent and a black smudge of burned and unburned powder particles may be evident along with fine metal particles. The velocity of the fine material quickly diminishes at greater distance, approximately 6 to 10 inches. Therefore, it does not strike the skin with enough force to stick. The scientist finds larger, heavier powder grains along with fair-sized particles of metal and grease, which adhere to the skin or surface of clothing.

These determinations can be crucial in discerning the distance between the shooter and the victim. If it can be established that the maximum distance for GSR deposit is greater than the length of the shooter, the absence of a powder pattern clearly indicates the two were not in proximity of physical contact when the shot was fired. However, other mitigating circumstances might indicate otherwise. Such an established fact eliminates suicide or an accidental discharge during a struggle.

GSR is subject to the elements of nature in that it may be washed off, or just disappears over a period of time. It all depends on where the person is and what he or she does. If a suspect is apprehended at the scene of the crime, he or she should be taken directly to the police department for processing. One way to conduct a preliminary test is to use a UV (ultraviolet) backlight. These lights, often seen around Halloween, fluoresce certain colors and show minute particles of dust that otherwise go unnoticed by the naked eye. Another reason GSR will fluoresce on the skin is because of the wavelength of the light. In white light, many particles on the skin are not visible. With a UV backlight, the particles of GSR stand out like someone who was using paint roller and the specs of paint are all over his or her arm.

When such a discovery is made, the detective or officer should photograph the area and carefully acquire a sample to be sent to the lab. Gloves, of course, are a must. One way of taking a sample is to use a clean swab, the end moistened with distilled water, to take a sample of the suspect area, trying to concentrate as much of the sample as possible on the end of the swab. In recent times, GSR tabs have been invented. They look like those little round stamps that kids use to ink happy faces all over themselves and their homework. The sample is collected by pressing the sticky end of the stamp on the suspect area. It is then properly labeled and sent to the lab. It might also prove useful to videotape and carefully document the process to avoid future complications in the courtroom.

Once a sample goes to the lab, the scientist, usually a toxicologist, uses the methods of chromatography and mass spectrometry to identify samples and try to match them up to samples from the crime scene. The instrumentation is so discriminating, it can match GSR to a batch or lot of bullets. The lot number appears on the packaging and can be traced to the bullet manufacturer. From there, it can be traced to the store where the bullets were shipped. If it happens to be a store in the vicinity of where the crime occurred, videotapes from the store can be subpoenaed and scanned for possible suspects.

If one has a job that deals with the handling of nitrates, he or she can be expected to test positive for GSR. Some of these jobs might include photographers, engravers, match workers, farmers, and landscapers who handle fertilizers. Other substances that may result in a positive test are bleaching agents, chemicals, cosmetics, explosives, tobacco, and urine.

The determination of the firing distance through test firings lies in the realm of the firearm identification expert. This person is not knowledgeable in matters of chemistry and instrumental analysis and thus should not be considered an expert in those fields in the courtroom. A certified forensic toxicologist or chemist can testify to the findings of GSR through the use of instrumentation in a lab setting.

If one is interested in becoming a forensic toxicologist, he or she needs a bachelor's degree in chemistry, toxicology, pharmacology, or chemical engineering. Additional on-the-job training, certification, and continuing education in the field are musts.

14

Confessions of a Con Man

The next morning, Chief Cahill received a call from the FBI. They were thoroughly impressed with the way Melinda and Fredric had handled the situation. It was the final piece to their puzzle. The FBI had enough information gathered and was ready to send in their decoy. Her name was Catherine Ferguson. She was one of the top agents in the field, and they had flown her in from California. Not only was she well versed in FBI tactics, she was also experienced in the world of wine. She was the perfect fit to act as an incognito wine critic and perhaps gain the confidence of Mike T.

As they wrapped up their meeting with the chief, Melinda's phone jingled. She looked at the incoming call and recognized the name Dr. Henry Rieders. "Here we go, guys. It's our man from National Medical Labs." She answered the call. "Good morning, my friend. Did you make it through the night?"

"Yes, actually it was a great night. I surprised some of my employees when they came in this morning. They were shocked that an old man like me could stay up that long. I'll probably go home at noon, but my work with your samples is done."

"What do you have, Dr. Rieders?"

"We have a match! I ran several gunshot residue samples through the ICPMS and all of them matched with a certain degree of scientific validity. The projectiles also matched with the sample from the body."

"That's great news," replied Melinda. "How did you make out with the prints?"

"I just received the results from my examiner. She says they match the bloody prints found on the knife. I think that's enough circumstantial evidence to warrant an arrest, wouldn't you say?"

"You can say that again," Melinda smiled. "You are the man, Doc. I'll be calling you in the near future to set up a dinner with you and your wife. We'll have a spectacular time, I promise."

"We'll look forward to it. Take care and I'll talk to you soon. Give my regards to the chief and Hassie. Bye."

"Well, we have a match!" exclaimed Melinda as she ended the call. "I can't believe we did it. Now we just have to wait for the feds to come and clean up the mess."

∽ ∾

FBI Agent Joe Corba was the head man in charge. He was a dapper kind of guy with a past filled with sting operations; he had apprehended some of the country's biggest criminals. His wits and intelligence were worthy of a MENSA membership and admission into the FBI hall of fame. The chief, Melinda, and Hassie greeted him with a hail-to-the-chief bow as he laughed and shook hands.

"You have set the groundwork for a prime opportunity to introduce our decoy. Catherine will be able to place Mike Tamasco in the optimum situation for inadvertently confessing about his dealings."

"How is that, Agent Corba?" asked the chief.

"Well, we know that right now is the highlight of the Chile grape season. We want Catherine to meet Mike Tamasco and lure him into the world of wine. From what we know about him, he's a suave guy who loves to show himself as the man, if you know what I mean. Catherine studied enology in California."

"What's that?" asked Melinda.

"It's the study of grapes for the purpose of cultivating and producing wine. She has her degree, and she would be able to entice Mike T. with dialogue concerning grapes and their value to the market. She's worked undercover before, and will be a great asset to our team. What I need you to do is sit back and let her work her magic. She's a stunning young lady, and there's no doubt in my mind that along with her proficiency in wine lingo, she'll snag Mike T. and acquire his trust."

"What exactly are your plans?" asked the chief.

Agent Corba sat back in his chair and sipped his coffee. "Catherine will make her way down to the warehouse and discuss the shipment of grapes from Chile. We set it up that she is a new employee of the Chadds Ford Winery. They brought her in to look at importing grapes to make a better, more sophisticated wine. We spoke to them, and they agreed to acknowledge her as an employee without going into further detail. They are being compensated for their cooperation as well. We'll have her wired for sound in addition to the existing taps we have in place. If all goes according to plan, she will have Mike T. talking about his business in no time. Of course, she'll use his machismo against him and engage him in conversations and possibly implicate him with the Mafia and his recent hits dealing with Adrian Belkoy and Phillipe Mindrago. If we can't nail him for shacking up with DeLuca, then we'll nail him for murder. It's been a long investigation and a lot of legwork, but we think it's time to bring this case to a close. If Mike T. is as big of a stud as he thinks he is, we can expose his weak spot with the alluring beauty of Catherine."

"When is all this scheduled to take place?" asked Chief Cahill.

"Tamasco is scheduled to be at the warehouse on Monday," replied Agent Corba. "Catherine will go down then and strike up a deal with him about purchasing a large quantity of grapes for the Chadds Ford Winery. With some flirting and a bit of luck, he'll take an interest in her and a friendship will ensue. I expect this testosterone-driven putz will want to take her out to dinner and try to wine and dine her. Then she can get him to talk. She has the knack. So what I need you to do is hang back for a little bit and let us do our thing. We appreciate everything your department has done for us, and we're willing to share the claim of the bust. It's not every day we work with the locals to apprehend a criminal."

"Sounds like a marvelous plan," said the chief. "What do you guys think?"

Melinda and Fredric nodded in agreement. At this point, they were satisfied with their progress in the case and were hoping the feds could lend them a helping hand. If their decoy could get Mike T. to talk about his personal business, it was a plus. They were now nearing closure on a case that had been plaguing the city. Kids at Roxborough High were facing graduation without two of the school's most prominent students. Melinda offered a handshake to Agent Corba. "We're with you all the way, and if you need anything, let us know. Otherwise, we'll wait for your call."

"We'll keep you informed of the situation. Actually, we'd like to invite you to sit in with us on our operation. You've done a lot of investigative work as well, and your presence could only assist us in capturing Tamasco. I'll be in touch."

As Agent Corba walked out, the chief, Melinda, and Fredric raised their coffee cups in an optimistic toast to the success of the plan.

⇛ ⇝

Monday morning was sultry as the humidity and haze lingered above the city. Melinda and Fredric were sitting with the feds in the back of a truck off Packer Avenue. It was a refrigerated truck, so the heat was not too intense. The equipment itself cost in the tens of thousands. Thank goodness for taxpayers, or none of this would be. They sat in the corner of the truck drinking coffee, while the feds ran tests on their gadgets and Catherine got wired up for her meeting with Mike T. She was a tall, slender brunette with skin as smooth as silk. She was a creature of beauty, and Melinda was the first to admit it. "Agent Ferguson, I must say you are stunning. I have no doubt you'll be successful."

"Thank you, uh...I'm sorry, what's your name?"

"I'm Detective Monsternick from the Philadelphia Police Department, and this my sidekick, Fredric Hassloch."

"A pleasure to meet you both," she returned with a handshake. "From what I've heard, this Tamasco guy is a real snake in the grass. However, not to worry, I've had a lot of training and I think I'll be able to help you bag him."

Now fully wired, Catherine put a few finishing touches on her makeup as the team gave her a salute. They waited for all traffic to subside before lifting the door to the truck. It was not your typical sight to see an attractive woman jumping down from the back of a truck to get into a car parked across the street.

She hopped into her new red Nissan Z and cruised off to the warehouse. She was on her way and all of the agents sent her a blessing, as they do on all undercover missions. Since the truck was situated off one of the side streets, the traffic was minimal. Once she turned onto Packer Avenue, the traffic picked up significantly. There were many thriving businesses on the street, and it was close to the Delaware River, where many of the city's imports were unloaded.

As Catherine approached the warehouse, men stopped and watched her. Being the typical construction workers, a few whistles could be heard and she turned to acknowledge them with a lovely smile that revealed perfect white teeth.

Once inside, she was immediately approached by none other than the con man himself. "Good morning, beautiful. I'm Mike Tamasco, but you can call me Mike T. Welcome to my warehouse. We're in the process of unloading a fresh shipment of grapes from Chile. We import grapes from around the world and are one of the largest suppliers on the East Coast. Can I pour you a glass of my personal Pinot Noir? It came from Oregon, which, aside from the Burgundy region of France, produces the best Pinot in the world."

"Actually, Washington State is finishing ahead of Oregon in the Pinot industry these days. The *terroir* is slowly changing and the warmer climate with foggy mornings is proving to provide much better growing conditions of the Pinot grape. But, yes, I would love to try a glass of your Pinot. My name is Ingrid McWhorter."

"Wow!" exclaimed Mike in a flirty tone. "You are educated in wine, I take it?"

"Why, yes I am," she responded. "I was just hired by Chadds Ford Winery to help improve some of their wines. I have an enology degree from Southern Cal."

"Yes, we provide all of the grapes to Chadds Ford. So they're having some problems with their wine?"

"There seems to be some trouble with the acidity and Ph balance. They're trying to even them out for a smoother taste on the palette. I think they haven't converted enough of the malic acid to lactic acid. That really is what gives wine that buttery feel. You know about maloactic fermentation, right?"

The guys in the van were laughing as they listened in. "He's hooked," claimed Agent Corba. "I knew she wouldn't let us down. Now we just need an invite to dinner and we're sitting pretty."

They continued to listen and Mike T. continued spilling his accomplishments about his business. His wine knowledge was so vast that he had been offered jobs from several wineries in France. He claimed that an Italian guy could not survive in France, so he had declined the offers. Then he started on his travels around the world to visit famous wineries and taste the grapes that he imported to his warehouse. His bragging eventually led to him to recount the story of his dining with one of the premier sommeliers in New Zealand and eating succulent ostrich dusted with crushed fennel seed and saffron threads served with a vintage shiraz.

Listening to his egotistical banter, Catherine already had him profiled. If his head were any bigger, it would need its own area code. She was tuned in to his personality and knew every single button to push. All the while, her dazzling smile kept him focused on her. It was obvious she had him wrapped around her finger. It was only a matter of time before he asked for a date. Sure enough, Mike T. obliged.

"So, what time should we meet for dinner?"

"What does that have to do with business?" she asked.

"It has everything to do with business. I need to hear more about your wine expertise. You could help my company sell more. I'm willing to pay you as a consultant, if you wish."

"How about we start with dinner? Can you get us reservations at one of the eminent wine spots in the city?"

"I can get us into Le Bec Fin. My friends run it. How about I meet you there around eight tonight?"

"I'll be there." Just then her phone buzzed. She answered and pretended to be speaking with someone from the winery. It was actually Agent Corba telling her to feign an emergency appointment and leave. She nodded as she hung up her phone. "I'm sorry, Mike, but I have to leave for an appointment. I'm looking forward to continuing our discussion tonight."

Mike T. watched as Catherine walked away, seeming like she was floating through the air.

Cupid had struck him straight in the loins and there was no doubt in his mind he had found himself a new girlfriend. He spent the rest of the day thinking about what stories to tell her at dinner that night so she would really be impressed. Little did he know that his bragging would be subject to serious scrutiny when it came time to take the stand in the courtroom.

 ∾   ∾

That evening found the two of them sitting at an open table at Le Bec Fin. The wine stewardess stopped by the table to personally introduce herself to Catherine. "Hello, my name is Andrea and I am the sommelier for the restaurant. Do you have any ideas on what you'd like tonight? I can recommend some great French wines to go with your dinner."

Catherine thought for a moment as she browsed the menu. Mike T. sat excitedly, for in his eyes, he was sitting with the sexiest girl in the restaurant. Catherine decided on an appetizer. "I'd like to start with the escargot."

"Excellent choice," noted Andrea. "I recommend you try one of our pinot blancs to cleanse the garlic and butter from your palette." She smiled, awaiting an answer.

"That sounds lovely, but I think I would prefer a sauvignon blanc from the Loire Valley. It's much brighter and crispier and will cut through the garlic and butter with a smoother edge."

Andrea blinked. Catherine was absolutely right, but Andrea did not want to admit defeat. "Yes ma'am, but the pinot blanc has a higher acidity and the citrus leaves the mouth with a dry coat."

"Yes, it is a dry coat, but it's a sharp contrast as opposed to the wine I chose. I'll start with a glass of the sauvignon blanc. You know what, bring the bottle, because I'll find a dish for dinner that also goes well with my choice. Thanks for your help."

Catherine gave her a snobby smile and she turned her attention to Mike T. who was gloating with pride. "She knows her stuff, doesn't she Andrea?"

Andrea smiled with pretended indifference. "Yes, she does, Mr. Tamasco. Should I bring you a glass of your favorite Bordeaux?"

"You sure can, sweetheart. I'd also like an order of the pâté. You have the best I've ever tasted. Ask Chef Jacques if he can add a little extra for my friend Ingrid here."

Andrea walked away in a huff. It was not her job to take food orders. She was considered a wine goddess and received all kinds of fancy write-ups in the local papers and magazines.

The surveillance team was parked around the corner in a garage laughing. For them, it was always a hoot to watch others with bogus intelligence made into a buffoonery extravaganza. Melinda and Fredric were truly enjoying themselves. All of them had had a meeting before Catherine left for her date. After listening to Mike T. brag about everything he could possibly spit out, Catherine would brag about a few stories of her own and try to top him. Being the one to instantly start the contest, Mike T. would try again and again to outstrip her. Eventually, the conversation would hopefully lead to gambling, sports, and money. Catherine was not interested in scripting any conversation because she felt if she made it up as she went along, it would seem more authentic. She did not want to spend time looking up at the ceiling jogging her memory. She had outsmarted more than one half-wit in her life and now was no different.

Throughout the evening, the pair exchanged many stories. As Catherine poured on the lies, more were returned by Mike T. If it was not a trip to the Alps and drinking wine in the gondola at 10,000 feet, it was riding in a gondola throughout Venice drinking fabulous vintages from the Piedmonts. Then stories about famous restaurants around the world and signature dishes that were prepared especially for them and their host started to surface. By the time dessert and cappuccino were served, the contest was far from over. It was a true matchup of the one-uppers. Mike T. seemed to be running out of stories as Catherine had instant returns to anything he said.

After the bill came, which was equivalent to a down payment on a small country house, the conversation subsided and Catherine thanked Mike T. for a wonderful evening. "It's only ten o'clock," he protested. "You're not getting off that easy. I'm having a wonderful time. How about we go for a ride in my limo and I'll take you on a tour of the city?"

"As long as we're back before I turn into a pumpkin, and my watch is broken."

Mike T. smiled and realized she was playing the game just as hard as he was. He phoned for his limo and they started their ride down South Street. Catherine looked on as the Goth kids roamed the streets with their punky hairdos, tattoos, body piercings, and dark, morbid makeup. "This kind of reminds me of some sections of Orange County," said Catherine. "These people are quite interesting. Most of them are harmless, or at least they were, until Columbine. It's amazing how people can kill on a whim and not even be fazed," she commented, cracking a diabolical grin at Mike T.

He ate it up and let out a, "Hmmm. Yeah, it's kind of interesting. I would imagine it would take a lot of guts to pull a trigger on someone."

"Yes, it would," returned Catherine. "I'd imagine it's a real adrenaline rush to shoot somebody. I mean, not kill them, but maybe, you know, shoot them in the leg or something. It must be a feeling of empowerment."

"You're not like other girls, are you, Ingrid?" he asked.

"Why, sure I am. I just tell the truth. All women have dreams. Most of them on the more passionate side, but causing trouble can be passionate in its own sense."

"You're so right, Ingrid." He started to inch his way over to her side of the limo.

She knew where he was headed, so she looked out the window and pointed out the Art Museum. "Hey, those are the steps Rocky ran up. Can we get out and run up them?"

"Sure, but you have on heels and a dress," he laughed.

"I can take my shoes off. Sure, I'll get a run in my stockings, but we won't be going out in public anymore tonight, will we?"

Mike T. let out a chuckle. "You are direct and to the point, aren't you?"

"Not always. C'mon, I'll race you."

The limo stopped in front of the steps. Both of them got out and Catherine took off her shoes. Mike T. decided to leave his shoes on but laid his sport coat on top of the limo. Catherine stood up and started running up the stairs. Mike T. yelled at her for cheating and chased her up the steps. She easily beat him and when he reached the top, he was out of breath. Catherine was laughing at him and calling a few names that most men resent, but actually resemble. He placed his arms around her waist and looked into her eyes. She returned the gaze of lust and gave him a quick peck on the lips. "The last one down is a rotten egg!"

She knew he would be too winded for the next 15 minutes to try anything in the limo. It was all coming together. The only thing left for the evening was to talk about sports, gambling, and money.

"Wow, you're a fast runner," he panted as they opened some bottled water. You must be in some pretty awesome shape to run like that without getting winded. I take it you go to the gym a lot."

"Yeah, I like to train for triathlons, and usually finish in the top ten. I have yet to win, but I can hang with the best of them. Hey, I want to drive by the sports area. I'd love to see the new stadiums."

"You're a sports fan, too?" he asked.

"I sure am. Football is first and foremost. Then college basketball. Everything else is for the worthless and weak."

By this point, Mike T. was in love. He had met the woman of his dreams. He started entertaining thoughts of a long-term relationship filled with trips to wineries around the world and sporting events at Lincoln Financial Field. His heart was booming, and he felt like a teenager at the Spring Fling, sneaking some alcohol and getting ready to make out in a car on a secluded hill overlooking the city. Catherine broke his thoughts.

"Who did you pick to win the Superbowl this year? No, wait, I bet you picked Seattle, didn't you?"

"I most certainly did not," he said, offended. "I might be a Philly boy, but I keep it in the state when it comes down to the nitty-gritty. The Steelers kicked ass this year, and so did my pool. I walked away with five grand."

"That's nothing. I walked away with ten."

Mike T. looked over at her and squinted his eyes in disbelief. "You're full of crap, Ingrid."

"No, I really did. I never said I laid down a bet to collect that money. You can do the math, Einstein."

"So, wait, like, you never laid down any bets, so you must have been collecting, right? You mean to tell me you're a bookie?"

"If that's what they call it on the East Coast. I prefer to call it entertainment expensing. Besides, a wine connoisseur is not on the *Forbes* top list of high incomes. It's expensive to live in California, especially in the Bay area. I need to exploit the public's weakness somehow."

"Oh, my God!" exclaimed Mike T. "This can't be. How the hell did I end up meeting a chick like you? You could not be more perfect for me. I run bets, too, and I make a few bucks. How long have you running bets?"

"About five years now. I was almost pinched last year by the LAPD, but my lawyer set me up with an alibi and that was the end of it. Most of those guys have better things to do, or at least that's what we pay them to think."

Mike T. shook his head incredulously. Fate had fallen into his lap this evening. He never thought he would see the day when he would meet his maker. Now his thoughts turned to marriage and a family business of wine and gambling. The two go hand in hand, just like him and his new love, Ingrid McWhorter. His excitement far exceeded his prudence for letting his guard down. "So, do you work for anyone out there or are you solo?"

"I don't have a boss, per se," she said. "But I do have to pay my membership dues to the brotherhood of protection. I don't get too big, and they don't get too pissed. Everyone is happy. How about you?"

"You've heard of DeLuca construction, right?"

"No, I'm afraid I haven't. What do they have to do with sports and gambling?"

"Nothing. But it's a front for my boss, Frank DeLuca, who pretty much runs this city. I bring in some big bucks for him. Hey, have you ever turned the screws on the sweaty little wimps who can't pay up?"

"I don't usually do it," she said nonchalantly. "I know some biker dudes who belong to the Hells Angels. They're cheap and enjoy the show. This way, my hands are clean."

She continued drawing him in. "One time, I had this high school punk who thought he was a big shot. His parents were rich, and his monthly allowance was in the thousands. He was a great customer until his parents went to jail for selling child pornography and he went belly-up. I still took his bets, and told him that if he didn't have the cash to back it up, not to bet. So this one time he bet fifteen thousand on the Belmont Stakes. He bet on this horse named Smarty Pants. Well, if you remember Smarty, you remember he broke his leg in the last turn—may his soul rest in a can of Alpo. This kid was bluffing and had nothing to give, and nothing coming to him. My boys left him in the bottom of a ravine for the animals. I felt bad for the loser for the first ten seconds, but then I cursed him up and down because I had to pay out of pocket for that."

"You think *he's* dumb," said Mike, "I have a similar story. There was this kid down in Roxborough. He was a little prick just like the kid you're talking about. He comes to me and tells me that these two jerks want to lay down a huge bet on the college basketball championship. It was close to fifteen grand, too. I tell him I'll take the bet and to come over with his buddies and we'll watch the game. They bet on Florida."

"That sucks for them," laughed Catherine. "That was one hell of a game. I couldn't believe it myself. I lost five on that one. I hate the Gators, but who knew Nova would become the Cinderella story? So did they pay out?"

"Of course not, so I had to meet up with them one day after school was out. Well, the one kid was walking back into the school when I turned the corner. I roughed him up and gave him a nice shiner. I couldn't find his buddy or I would have socked him, too. That money goes to DeLuca, and that was coming out of my pocket. I went to him and explained the situation. He said that if I whacked the two kids, I wouldn't have to pay him. Now, fifteen grand is a drop in the bucket for me, but I like the challenge.

"So I call this kid and tell him to meet me down at the Wissahickon in Fairmount Park. When I told him to bring his sorry-ass friend, he made up some bullshit excuse that his grandmother died and he would be coming alone. I told him that I was mad that day I socked him and apologized for smacking him up. I just wanted to talk about his options, I said. Maybe he could do me a favor and we would be even. Of course, the moron shows up. We go for a walk off the beaten path. He was nervous as hell. I tell him to calm down because I wasn't going to hurt him. Ten minutes later, he started to relax. That's when I took out my .45 and made him get down on his knees. He begged for his life and offered me anything I wanted. I said, 'how about my fifteen grand?'"

Catherine was pretending to be enthralled by his story and clinging to every word. Mike T. could tell she was his kind and they were going to live happily ever after. She was as demented as he was, and hopefully this was the beginning of a budding relationship that would end in holy matrimony.

Mike T. continued his story. "So this kid tells me he can't come up with the money and his buddy had to leave town because his grandmother died. Likely excuse, right? I told him to save his story for the pearly gates and popped one in his head execution-style. I like the concave mark left in the skull when you pop a hollow point from about four inches."

"What happened to the kid who went to the funeral?" inquired Catherine.

"I thought about him. The first kid wasn't found for like a month and the police were searching everywhere and couldn't come up with anything. I've been in this business a long time, and I know how to cover my tracks."

"I could learn a lot from you, Michael," she said shyly.

"Call me Mike T. like everyone else. Michael sounds too formal. Anyways, the investigation slows down and nothing is showing up in the papers or on TV. I start tailing this other kid for a day or two. I still had a deal to keep with DeLuca, so the only choice I had was to take him out, too. He was at home alone one day. I guess he was skipping school or something, because it was during the week. I rented a car, and of course I put on a wig, fake mustache, and all that stuff. I knocked on the door and when he answered, I pretended like I was looking for a friend of mine. After he said he didn't know the guy, he looked at me funny. I saw it click in his head; he recognized me. After all my hard work, he saw right through it. So he runs into the kitchen and grabs a knife. I had my gun, but I couldn't resist the challenge. Here was this punk, who was pretty well put together, by the way, thinking he was going to take me out. He starts jabbing and swinging at me with the knife. You're not going to believe this, but he takes a swing and bangs his hand on the stove. He yells out a few cuss words and almost drops the knife. That was my chance, so I kicked him in the chest and knocked him down. He took one more swing at me and cut my thumb. It stung like hell and the blood was dripping. Now I was pissed. I grabbed a chair and knocked him down. I took the knife and made a nice clean cut through the neck. I stepped back because that stuff goes everywhere. He's gasping for breath and as he's sitting there on his knees, I popped one in the back of his head. This way, when the cops came to check it out, they would suspect a serial killer because I shot them both the same way. They were also two high school kids. You should've seen the papers. Everyone was afraid a serial killer was running loose on the streets of Philadelphia. I laid low for a while, cleared my name with the boss, and here I am."

"That is one impressive story. Shall I call you master?" grinned Catherine.

"You may," he said as he raised his eyebrows.

The limo was now returning to their starting point. As they pulled into the parking lot, Mike T. offered Catherine a nightcap back at his place, and whatever else she had in mind. Her face showed one thing, while her mind was disgusted with this abominable creature, who should be placed under a rock and left there to rot.

"I would love to, sweetie, but I have to be in early tomorrow. How about we finish this on Friday? I have to do some traveling for the next few days and won't be back until then. How about you come to my place and I'll cook dinner for us? I'll supply the wine, too. Maybe afterward, we can catch a Phillies game, and then who knows what. It'll be the weekend, you know," she said suggestively, leaning over to give him a short but juicy kiss. She still kept her distance so he did not have a chance to embrace her. He might have felt out the wiretap running under her bra. She jumped out of the car and winked at him with a lusty smile that foreshadowed good things to come.

"We got it," said one of the agents as he took off his headphones. "What do you think, boss?"

Agent Corba gave a round of high fives to everyone. "We did it. Let's follow him back to his house and bust him there. I can't believe he was this stupid. This was easier than I thought. I was expecting a couple of weeks before she cracked him. Melinda, can you call the judge and get a warrant for us?"

"I sure can. I'll call him right now and we can pick it up on the way over there."

Mike T. was brushing his teeth when the doorbell rang. He was hoping it was Catherine coming to claim her nightcap. When he opened the door, FBI agents busted in and started yelling. "Get down on the ground! Put your hands behind your back!"

Mike T. was at a disadvantage because he was in his boxer shorts. He couldn't believe what was happening. It was almost like a nightmare he could not wake up from. Agent Corba stepped in and dangled his handcuffs in front of him. "Michael Tamasco, you have the right to remain silent. Anything you say can and will be used against you in a court of law. You have the right to an attorney. If you cannot afford one, one will be provided for you by the City of Brotherly Love. Do you understand these rights as I have read them?"

"What the hell is this? I haven't done anything wrong! What are you booking me on?"

"We have a warrant for your arrest in the murders of Adrian Belkoy and Phillipe Mindrago. Put on some clothes, and then we're headed to the station."

Mike T. snorted and huffed as the officers accompanied him up to his bedroom to get dressed for his visit to the precinct.

15

Justice for Jocks

Mike T. sat with his lawyer, Stephen Petrozzo, at a brown walnut table. On the other side resided Agent Corba and Renee Merion, the federal prosecutor. The course of Mike's life depended on the upcoming trial, for he had been a bad boy over the last 10 years. Not only was he the number-one hitman for the infamous Philly mobster Frank DeLuca, but he was also the suspect in the deaths of two young promising lads who would never be able to bring their dreams to life. It was going to be a long day. Petrozzo opened the floor.

"You know you won't be able to pin all of this on him, so what are you offering?"

"*Au contraire, mon frère*. Let's see," said Renee, "we have five months of wiretaps that can place your client and Mr. DeLuca in some rather precarious situations. Notwithstanding the current double-homicide charge, we have racketeering, money counterfeiting, and many other illegal activities that your client was privy to with DeLuca. We would like to place both in the prison system. Under the circumstances, there is not much we can offer without the testimony of your client."

Petrozzo sat for a moment scribbling down notes on a yellow legal pad. Mike T. sat calmly with a smirk on his face. He had been in these situations before, and it was only a matter of time before someone would be compensated for false testimony or evidence was misplaced, never to be seen again. As Petrozzo flipped through some of the documentation that had accumulated over the last several months, he leaned over to whisper in Mike T.'s ear. "There's a lot of implicating evidence against you. Some of it I might be able to prove circumstantial, but some of it seems irrefutable. Let's see what they have for us and go from there. You better stiffen up, too, because with that wise-ass grin on your face, you won't go anywhere but straight to jail and you know what will happen there." He sat up and addressed the prosecutor. "Ms. Merion, what are you offering?"

"Fifty to life if he supplies testimony about Frank DeLuca and helps us put him behind bars. We might be able to waive the death penalty, but we need confirmation from the upper echelon of the justice system. We are more than happy to change his name and place him in an out-of-state facility that will provide him all the luxuries the state has to offer." She smiled at Mike T. as he sat there staring at the floor, his smirk now gone. It was inevitable that he was headed to prison and possibly death. If anyone found out where he was placed, a hit on his life was assured.

"May I have a moment with my client, please?" asked Petrozzo.

"You sure can," replied Renee. She and Agent Corba stepped out of the room. It was more than probable that the room was bugged, so Petrozzo spoke in a whisper, covering his mouth.

"You're in deep this time, Mike. You got nailed for killing two kids. They weren't chumps, either. The public wants to hang you, and if they don't get you first, DeLuca will be looking for you. Taking the death penalty off the table isn't the best offer, but at least you might get parole in fifty. I can't bargain much here. I love you like a son, buddy, but you really screwed up this time. I know some people in the upper echelon as well, so we'll ask for a day to sleep on it and get back to them tomorrow."

"What if they don't give us a day?" inquired Mike T.

"I'll work it out. Just keep your mouth shut and look sorry for yourself."

Petrozzo signaled for them to come back into the room. "We think you have a legitimate offer and we would please request twenty-four hours to think about it. I promise you we will be in contact tomorrow."

"That's the offer, Mr. Petrozzo," said Renee. "Take it or gamble your luck with the jury. This is one of the highest-profile cases of the last decade. I'll give you twelve hours to make your decision. If I don't hear from you by ten this evening, all bets are off. Good day."

As Mike T. was escorted back to his jail cell, Petrozzo offered a little encouragement. "Look, buddy, I'll do what I can for you. I'll try my utmost to get you into a secluded jail somewhere nobody can find you. Start practicing your tears now and muster some remorse out of the black hole inside you. We're going down, but we're going down swinging. See ya later, Mike."

Nine hours later, Renee's cell phone rang, and the name Stephen Petrozzo came up on the display. She knew he was sharp and had defeated many prosecutors. His working knowledge of the law was equivalent to that of inspector number twelve at the Fruit of the Loom factory. Nothing went past him. Renee fielded his call in a sweet voice. "Yes, Mr. Petrozzo, what have you decided?"

"Ms. Merion, I must say that I've been in many courtrooms and beaten more than my share of prosecutors. But I've spoken to my superiors, and they have advised me to comply with your offer. I'll be in touch soon to set up a meeting." He sounded defeated as he hung up the phone.

Renee smiled as she sat back in her chair and placed her hands behind her head. The forthcoming trial was going to be utter chaos and the media unrelenting. She smiled again as she thought about the chance to make a name for herself by defeating the infamous Stephen Petrozzo.

For weeks the buzz had been building in Philadelphia as well as across the country. Frank DeLuca, the ubiquitous Mafia icon, had been indicted by a grand jury for racketeering, counterfeiting money, tax evasion, and other crimes. Adding more fuel to the fire, Michael Tamasco was awaiting trial for the murders of Adrian Belkoy and Phillipe Mindrago. Court TV arrived the week before the trial started to set the backdrop. They interviewed the citizens of Philadelphia and the reviews were mixed. Some despised DeLuca while others praised him for his hearty philanthropy throughout the city. On the other hand, Mike T. received no sympathy except from a ditzy blonde who appeared on the afternoon news. She claimed that Mike T. was a cutie and she would forgive him and wait for him until he got out of jail. A pen-pal relationship was in his near future. Other than that, the residents were hoping for the death penalty. The city was quite unforgiving when it came to personal business. There were two things not to be messed with: their sports teams and children who represented the All-Americans of the country.

The DeLuca trial started first with opening statements that dragged on for three hours. It was several days before witnesses were called to the stand. The experts ran far and wide to testify for and against the hardened mobster, DeLuca. Conflicting opinions stirred controversy among the jurors and dissension in the ranks was evident. It was exactly what the defense was hoping for. A hung jury meant a victory for them because DeLuca could walk free and the lawyers would receive their annual bonus for keeping him out of jail. The courtroom was closed to all media, including Court TV. All of them waited anxiously every day for court to be adjourned, so they could bombard the players as they left the courthouse. The reporters were lined up along the steps while the cameras rolled. Cameras flashed as microphones were shoved into DeLuca's face and those of his attorneys. Several other members followed in tow, awaiting their five seconds of fame. All of them had been informed to not say a word, and DeLuca would see to it that any who did would be riding in a hearse the next day. Obligingly, they obeyed their orders and continued to walk with him avoiding the reporters and keeping a impassive face for the cameras.

By the end of the week, the trial had been sensationalized enough to make news around the country. Larry King and Oprah were looking for consultants to have on their shows. Contradictory arguments put a hefty dent in the prosecution's case, but they had not yet unleashed the wiretap tapes. There was also room for an eyewitness or two that would hopefully be convincing enough to devastate all other disagreeing arguments and send DeLuca to jail.

As the case picked up speed, it came time for Mike T. to testify. He had agreed to testify against DeLuca in exchange for a lesser sentence, thus avoiding the death penalty. DeLuca would not serve too much time if convicted. He would probably be sent to a minimum-security prison with frequent visitors and meals shipped in from the outside, somewhere in the vicinity of where Martha Stewart had stayed during her short exile from the kitchen.

Mike T. was hoping for the same situation. He was not as big as DeLuca, so he could only hope to find a small prison somewhere to house him for his sentence. DeLuca was not aware that Mike T. was going to testify against him. The prosecution was holding him as their smoking gun, along with the tapes. The trial was now entering the third week, and the prosecution was ready for the unveiling of five months of long nights and lots of stale coffee. Many conversations had been recorded and a lot of illegal activity revealed. The defense attorney for DeLuca tried to refute the tapes as inadmissible because the correct procedures had not been followed. Acquiring a wiretap in Pennsylvania is one of the more difficult tasks for police officers. However, when the FBI is involved, those permission slips are granted more easily. The judge overruled and let the tapes stand.

It was discovered that Mike T. and DeLuca had made numerous arrangements for drugs, stolen goods, and counterfeit money to be sold along the eastern seaboard. A specialist from the FBI also testified that the voices were those of Frank DeLuca and Michael Tamasco. He used a voice recognition system and explained to the jury how the wavelengths of the tapes matched the voice samples from DeLuca and Mike T. The defense attorney tried to shoot holes in the testimony, to no avail. It was more than circumstantial evidence at this point.

∽ ∾

The next day, Mike T. took the stand and placed his hand on the Bible as he was sworn in. He agreed to tell the truth and sat confidently in his chair. As the prosecution probed into his conversations with DeLuca, he answered without flinching. It was obvious the prosecutor had spent time preparing him for the trial. Renee would ask questions, and he would answer them without batting an eye. Not once did he make eye contact with DeLuca. Mike T. knew he was the rat in the courtroom, and if it saved his life, then all the more reason to testify.

DeLuca's defense attorney tried to shake him up by asking if he had been offered a deal in exchange for his testimony, and if that were true, how did the jury know that he did not make up some of the information to sway the decision of the jury? It was objected and Mike T. did not have to answer the question. The jury knew that he was going to be facing a jury of his own in the upcoming weeks. His answers were truthful because they matched the answers that were recorded on the wiretaps. He had done his job and the prosecution rested its case. The defense did not have much to counteract all of the testimony, so they tried a few witnesses of their own whose vague testimony proved to be more of a hindrance than help. They gave different stories that exposed them as liars. Instead of charging them with perjury, the judge let the jury use the testimony as one more reason that DeLuca was guilty. Court was dismissed and the jury headed into deliberations.

It took only two hours before the decision was returned. Frank DeLuca was found guilty of racketeering, illegal gambling, counterfeiting money, drug smuggling, and tax evasion. Because there was no solid evidence that connected DeLuca with the murders of the boys, those charges did not stick. He was sentenced to 30 years with the possibility of parole in 20. It was a small victory for his team, because there would be appeals and his parole would probably be pushed up a few years. DeLuca was happy with the turnout because he was headed to a cushy prison where he could have lots of free time and the criminals were not the hardcore rapists, murderers, and serial killers that plagued the tougher correction facilities of the state.

∽   ∾

The following Monday, the trial of the state of Pennsylvania vs. Michael Tamasco began. Renee Merion already had one notch on her belt and it was time for another. She was on a roll and on her way to becoming a seasoned prosecutor. The jury selection was an even mix of men and women. Some were blue collar while the others were members of Fortune 500 companies. It was an exhaustive process and neither side was happy with the turnout—the defense wanted more blue collar and the prosecution wanted more white collar—therefore it was fair.

The courtroom was filled with a vast array of people ranging from parents to students to neighbors, to wine makers.

Seated at the defense table, Mike T. tried his hardest to look sorry for himself. Several times he was seen staring off into space as if he had no clue where he was. Petrozzo nudged him a couple times under the table to bring him back to reality. As witnesses took the stand and provided testimony, he looked at them apathetically like they were of no importance to his trial. Some of the most damaging testimony came from Dr. DiGregorio, Arthur Ying, and Dr. Henry Rieders.

Dr. DiGregorio was first to take the stand. The first 10 minutes were used to establish his background. He had been the Medical Examiner for the city of Philadelphia for 30 years. He had performed over 3,000 autopsies and was part of several elite organizations for pathologists. At one time he served as the president of the American Academy of Forensic Science, the Society of Pathology, and vice president of the American Board of Health. His staunch reputation labeled him a pillar in the medical community.

Renee was aware that members of the victims' families were in the courtroom, and thus she was not too gung-ho about exposing the atrocities of the crime. Nevertheless, she needed to expose Mike T. for the ruthless killer he was. Her line of questioning was direct and established the insidious violence upon the two boys.

"Tell me, Dr. DiGregorio, when you examined the body of Phillipe Mindrago, what did you find?"

"The body was very badly decomposed. It was in the latter stages of decomposition. I collected several samples of flies and other fauna that were on the body. Detective Monsternick took them to Dr. Blass at National Medical Labs. It was too late to tell if there had been any trauma to the body, so following procedure, I took X-rays of the entire body."

"Can you show us those X-rays and explain what you discovered?"

"Yes, the first X-ray is of the head."

"Let it be noted," said Renee, "this is exhibit number sixty-two. Go ahead, Doctor."

"As you can see, there is a slight concave opening in the left parietal bone. That is the bone at the top portion of the head. This is where the projectile entered the head. The projectile itself is not visible in the X-ray, because it made a clear path through the skull and out of the front, which is my next X-ray."

Renee interrupted again. "Please let it be noted that this X-ray is exhibit number sixty-three."

"This X-ray is the front view of the skull," Dr. DiGregorio continued. "There is an exit wound on the left zygomatic bone, otherwise known as the cheek area. Based on my knowledge of gunshot wounds, and the research I have completed in the field, I would say with a certain degree of scientific validity that the projectile found underneath the body caused this wound. I could not autopsy the brain because of the decomposition."

"So a projectile was found underneath the body?" asked Renee.

"Yes. According to Detective Melinda Monsternick and Detective Fredric Hassloch, a projectile in the shape of a bullet was found. I cannot speak to that because that is not my field. I do know it was photographed and documented as evidence."

The courtroom was so silent you could hear a mouse passing gas. They sat transfixed as Dr. DiGregorio continued his testimony.

"What about the second body, Doctor? What were your findings?"

"The body was identified as that of Adrian Belkoy. Based on the rigor mortis and livor mortis, I would estimate he had been deceased for approximately five hours, give or take an hour."

"How was he killed, Dr. DiGregorio?"

"He died from exsanguination, which is loss of blood. There is a clear cut through the jugular, which released the blood. I also found some trauma to the chest, which was antemortem."

"Can you explain that for the court, Doctor?"

"Yes, there was a large bruise on the breastplate of the victim. Upon further investigation, I found hemorrhaging around the area, which is indicative of an antemortem bruise, which means it happened before he expired."

"There was also a gunshot wound involved, correct?"

"Correct," returned Dr. DiGregorio. "The entrance wound was in the left parietal bone, same as the other victim. It was also a direct-contact gunshot wound, because there was evidence of starring."

"Can you explain that to the court, Dr. DiGregorio?"

Dr. DiGregorio enlightened the court about gunshot wounds, the difference between distances from the shooter and the body, and how samples were collected to send to the lab. Many members of the courtroom had to excuse themselves, for a picture was developing of how Mike T. had killed the two kids. After he finished his extensive but enlightening lecture, Renee had one more question. "Could you explain which came first, the cut through the jugular vein or the gunshot wound?"

"The cut through the jugular was first. There was no reason for the gunshot. After a major vein or artery is cut in the throat, the victim has approximately ten seconds before all the blood drains out, and the person expires. There was no evidence of hemorrhaging around the gunshot wound, which leads me to believe the blood had already vacated the body."

"Dr. DiGregorio, do you mean to tell me that death was not instantaneous? Is it possible that he suffered for a short interval of time?"

"Yes."

The courtroom let out gasps and whispers could be heard as Renee turned it over to Mr. Petrozzo. He made a weak attempt to refute the evidence. Dr. DiGregorio's credentials were impeccable, so he attacked his methods. "Dr. DiGregorio, how can you really tell that he suffered?" Do you have any research to substantiate that?"

"Why, yes, there are several published papers in the professional medical literature about crimes of this nature and the effect on the body."

"Well, have you slit someone's throat and then tried to ask them how long they can stay alive? I just don't see how that kind of testing could be considered valid in the medical field."

"Mr. Petrozzo, you would be amazed what kinds of studies have been performed in the name of medicine. We would never think of any such terror to inflict upon another human for a study. However, many tests on animals were performed under the most scientific conditions to acquire the appropriate information regarding the matter. At the mercy of the court, I would rather not discuss it here due to the sensitivity of the case."

"No more questions, Your Honor," Petrozzo huffed as he returned to his seat. Renee had won this battle but the war was not over.

The next person to testify was Dr. Arthur Ying. He pledged his oath to tell the whole truth and nothing but the truth so help him God. He sat attentively while Renee provided his background and pertinence to the case.

"So, tell me," inquired Renee, "We know that you are a certified forensic biologist specializing in the area of DNA. Can you give a brief synopsis of DNA importance in crime scene investigations?"

"I sure can," he replied with a wide grin. "DNA is what makes us, well, us. It determines our eye color, hair color, the development of organs, the shape of the body, and so on. Our genetics are passed on through our DNA. That means that hereditary traits are in there as well. There are two kinds of DNA; one is nuclear DNA, which is a fifty-fifty mix from your mother and father. This DNA resides in the nucleus of the cell. The other type of DNA is called mitochondrial DNA. This is from the mother only. After an egg is fertilized, the tail of the sperm falls off, losing the mito-chondrial DNA of the father. DNA is important because we can separate samples and determine who belongs to what samples. Then once we match the DNA to the appropriate owner, we can discriminate between the victim and perpetrator."

"What kind of DNA did you test?"

"We tested mitochondrial DNA."

"What is it that makes DNA unique to other individuals?"

"It is the so-called junk DNA that is examined. Most of our DNA is the same. That's why we have eyes, a nose, ears, and a mouth. The junk DNA is composed of small structural differences in the repeats of amino acid sequences that can be seen through molecular examination. These are called Standard Tandem Repeats, or STRs. All you need to know about amino acids is that there are four main acids involved in DNA con-struction. They are adenine, cytosine, guanine, and thymine. As these four bond together, they form strands that eventually are developed into the double helix."

"What is the importance of this double helix?" asked Renee.

"What we do is separate the double helix strands. In this particular case, we used a process called PCR. That stands for Polymerase Chain Reaction. Before I confuse the jury any more, let me explain. We use temperature changes to help us out. The DNA is exposed to different temperatures, and that helps us in three ways. One, it separates the DNA into two strands." Dr. Ying showed an illustration to the jury. "After DNA is separated, the temperature changes again. Now it can reproduce the same exact structure of the single strand of DNA. During this phase, we also incorporate the use of chemical helpers to assist in the process. The temperature is then changed again and the DNA will bind together to make another copy of the double helix. This process runs through thirty-two cycles. At the end, we have upwards of a billion copies. It can go higher, but that is usually substantial enough for us scientifically.

"So, what did you test and what did you find?"

"I tested samples from two separate bodies. The first was a tissue sample, which was identified as belonging to one Phillipe Mindrago. The other specimen I tested was a tissue and blood sample, which I identified as belonging to one Adrian Belkoy. I also tested another sample that was found on the bloody knife at the crime scene."

"And who did that sample belong to?"

"One profile belonged to Adrian Belkoy and the other to Michael Tamasco."

"Is Michael Tamasco in this room right now?" asked Renee.

"Yes, that's him at the defendant table in the gray suit with the white shirt and blue tie."

"Let it be noted that Dr. Ying has identified Michael Tamasco as the donor of the DNA sample found on the knife."

Adrian's mother could be heard sobbing in the courtroom as Dr. Ying continued his testimony. He knew she was in pain, so he tried to keep it as simple as possible.

"What are the odds of someone else having the same DNA?" inquired Renee.

"I would have to say that with today's technology, one in one trillion."

After Dr. Ying gave some more explanations and proved with a certain degree of scientific validity that the samples matched the two boys, Stephen Petrozzo had his opportunity to cross-examine. He started off with an attack on Ying's credentials. "Dr. Ying, you have been working for National Medical Labs for five years now, correct?"

"Yes."

"Where were you before that?"

"I was finishing up my doctorate's degree in forensic biology."

"Where did you obtain your degree, Dr. Ying?" asked Petrozzo.

"I went to Louisiana State."

"Are they an accredited university?"

"At the time, they were awaiting their certification. They did receive it from the American Academy of Forensic Science the year after I graduated."

"Well," gestured Mr. Petrozzo, "if the university was not accredited at the time of your matriculation, how do we know that you received the best training possible?"

"Objection, Your Honor," interjected Renee. "This has nothing to do with the case."

"I'm simply trying to establish a background here," returned Petrozzo. "Getting back to Dr. Ying, how many cases have you worked on over the last five years?"

"Probably over a hundred."

"Have all of them involved DNA testing for homicide investigations?"

"No," responded Dr. Ying. "Some were for paternity cases, civil cases, and some rape cases."

"Then how many cases have related to homicide investigations?"

"I would have to say approximately twenty-five."

"Do you think that's enough to say that you are an expert?"

"I would say so," stated Dr. Ying. "It doesn't matter what kind of investigation you have. It matters that the instrumentation is run properly and the results are identified. In all of my cases, I have correctly identified my samples. They were even double-checked by the supervisor, who has been there for thirty years now."

"One more question, Dr. Ying. How did you acquire a DNA sample from Mr. Tamasco when he was never asked to submit one?"

"Well, that's one for the FBI and police. I do believe they picked an empty glass of wine out of the garbage that he had consumed at his place of employment. Once it's in the trash, it becomes public domain. I believe all of the proper paperwork was submitted as well."

By this point the prosecution was ahead of the game. The experts provided solid testimony and explanations that even the most lay of lay persons could understand. The trial was picked up by Court TV, and the media was giving the trial more attention since Mike T. had just testified against DeLuca and helped send him to jail.

In the next few days, several other people testified about different pieces of forensic evidence and the importance to the trial. Dr. Blass provided excellent testimony that explained the process of decomposition and the possible hours of the crime. It was up to Mike T. to provide an alibi for those hours. The time estimation of death was within 12 hours, so Petrozzo would need to present witnesses that would corroborate Mike T.'s alibi. He brought in Lenny, Mike T.'s partner from the wine warehouse to say they'd been together at Lenny's sharing some drinks. When the judge reminded Lenny that he was under oath and could be prosecuted for perjury, he quickly changed his story to say he'd only spent a few hours with Mike T., and then Mike T. had left by himself, giving him time to commit the murder of Phillipe.

Mike Ragney appeared as the fingerprint expert and provided a fascinating account of fingerprints. The courtroom hung on his every word because he was a good-looking guy and a terrific speaker. One would think that the Pope was giving an oratory on world peace. The only thing missing was applause at the end and a request for an encore. He buried Mike T. further into the ground by testifying that the prints on the gun matched Mike T.'s.

Melinda Monsternick, Fredric Hassloch, and several FBI agents testified about the tape recordings and how they legally acquired their information. Catherine Ferguson also testified to coincide with the testimony of her colleagues. Some of the students also testified about the

Drater Club and the underground hoopla of illegal gambling going on at Roxborough High.

When Carmine DiGuiseppe took the stand, he fingered Mike T. as the one who had taken the bet from the two boys. He had no knowledge about the murders, but Judge Pagan made sure to instill the fear of God into him by promising a horrible life in the prison system if he did not clean up his act.

One of the last people to testify was Dr. Rieders. As an eminent figure in the forensic field, he had testified in many criminal cases. It was one thing to sit and listen to his stories while he was explaining scientific terms, but in the courtroom, he spoke magnificently about his field. His testimony about the results of his tests on the gun used to kill both boys was spectacular. He brought it down to the level of the courtroom. If his day job did not work out for him, he could go into teaching, because it was one of the most interesting testimonies of the trial. He explained the difference between ballistics and firearm identification, how he identified the projectiles from both bodies and matched them to the gun owned by Mike T., and the identification of gunshot residue through the use of inductive coupled plasma mass spectrometry. Petrozzo tried his hardest to discredit the testimony, but failed miserably. The witnesses for the defense proved to be as invaluable as the ones who testified in the DeLuca trial, for any trained mobster knows about telling the truth and where it lands them—a coffin. Petrozzo, not wanting to admit defeat, took the only choice he had left, he rest his case.

The jury went into deliberations. They were sequestered by this point, because the media was on them like stink on dog dirt. It was requested by Renee on the second day of the trial, because she knew if she left them open to the media, someone would talk and that would result in a mistrial. This way, they had to stay together and that would force them to talk more and focus on the case. Some of the jurors were angered, while others looked at it as a vacation from the real world. The testimony against Mike T. was quite damaging, but some of the jurors were not convinced of his guilt. Two of them had a slight doubt about Mike T. After hearing all of the great things he had done for the community, they could not see how this man could commit murder. They thought the gunshot residue tests were bogus, and wondered if someone had planted the evidence at the crime scene to set him up. It also came out in the witness testimony that Mike T. had enemies as well as friends. There were some what-ifs for the jury and they could not make up their minds. The jury forewoman, Jane Hamley, made everyone stay up the entire night. There were lots of arguments, and before they knew it, it was time to return to the courtroom. If they could not agree, it might result in a hung jury, which meant freedom for Mike T.

Both parties filed into the standing-room-only courtroom. Melinda and Fredric were there, as well as Agent Corba and Catherine Ferguson. It was a tense moment as everyone waited to hear from Jane Hamley.

"Have you reached a verdict?" asked the judge.

"Yes, we have, Your Honor. We the jury find the defendant, Michael Tamasco, guilty on both charges of homicide."

The courtroom erupted into cheers and jeers. Mike T. stood up and kicked the table, uttering some cuss words under his breath. The judge hammered his gavel on the bench and demanded order. Justice was finally served for two kids who had never had a chance. As all of the commotion settled, the judge regained order in the room. "We will adjourn for now and reconvene in two weeks' time for sentencing." He slammed down his gavel and court was dismissed.

People outside were celebrating and singing praises. Others were holding up signs that asked for the death penalty. Melinda and Fredric were standing with Agent Corba and Renee Merion.

"Fantastic job, Ms. Merion," offered Melinda. "I'm very impressed with you. You nailed two of our city's most wanted and sent them on their way. It's not the end, but it's a great beginning."

"I second that," added Agent Corba. "You did a stellar job proving our case. I know we agreed to make a deal with them about the death penalty, but it's possible the judge could overrule and sentence him to the death penalty, right?"

"Technically, he could, but I've spoken to him. He seems to be leaning in the direction of taking the death penalty off the table, but he's been known to change his mind—in which case, Mike T. is done."

Two weeks later, court reconvened with just Mike T., his lawyer, Renee Merion, and Judge Pagan. The judge looked sternly at Mike T. over his glasses. "Michael Tamasco, given the circumstances of this court case and the testimony I have heard, I am hereby sentencing you to two life terms in a maximum-security prison that I will designate. I am aware of the mitigating circumstances, and instead of giving you the death penalty, your life will be spared, but you will have to serve two life sentences without parole. Court is adjourned."

Mike T. looked incredulously at his lawyer, Stephen Petrozzo, who stood with his mouth agape. "Judge Pagan, my client was promised a chance for parole. This just isn't fair!"

Judge Pagan removed his glasses and stroked his finely trimmed goatee. "You know what? You're right, Councilman Petrozzo, it's not fair to go back on promises that were made, but it's also not fair that two boys are dead because they couldn't keep a promise, either." With that, Judge Pagan exited the courtroom.

Two days later, Melinda and Fredric sat in the chief's office. He was pleased with how the whole situation had turned out and congratulated them both on a job well done. "Hassie, I would like to give you a most sincere congratulations on your accomplishments. You have now earned your batwings." With that he handed him a pin of the batman insignia. It was not one to wear on the uniform, but rather one to file away in the personal accomplishment folder. A rookie was now part of the permanent team.

Fredric thanked the chief for his blessing. "I'm so happy to have made it through my initiation and am now considered one of the department. I look forward to assisting the city in many future cases."

Just then the door opened. It was one of the officers. "We just received a call that a body's been found in a Center City apartment with an execution-style gunshot to the head. The place is a mess."

The three looked at each other in surprise. Melinda was the first to open her mouth. "Could it be a copycat, or was Mike T. really set up and we have a serial killer on our hands?"

Further Reading

Bard, Mike. *Crime Scene Evidence: A Guide to Recovery and Collection of Evidence.* California: Staggs Publishing, 2001.

Bevel, Tom, and Ross Gardner. *Bloodstain Pattern Analysis: With an Introduction to Crime Scene Reconstruction, Second Edition.* New York: CRC Press, 2001.

Byrd, Jason. *Forensic Anthropology: The Utility of Arthropods in Legal Investigations.* New York: CRC Press, 2000.

Committee on the Judiciary. "Issues Surrounding the Use of Polygraphs." Hawaii: University Press of the Pacific, 2005.

DiMaio, Vincent. *Forensic Pathology, Second Edition.* New York: CRC Press, 2001.

———. *Gunshot Wounds: Practical Aspects of Firearms, Ballistics, and Forensic Techniques, Second Edition.* New York: CRC Press, 1998.

Dix, Jay. *Color Atlas of Forensic Pathology.* New York: CRC Press, 1999.

———. *Guide to Forensic Pathology.* New York: CRC Press, 1998.

———. *Time of Death, Decomposition, and Identification.* New York: CRC Press, 2000.

Dodd, Malcolm. *Terminal Ballistics: A Text and Atlas of Gunshot Wounds.* New York: CRC Press, 2005.

Dolinak, David, Evan Matshes, and Emma Lew. *Forensic Pathology: Principles and Practice.* New York: Academic Press, 2005.

Douglas, John, Anne Burgess, Allen Burgess, and Robert Ressler. *Crime Classification Manual.* San Francisco: Jossey Bass, 1992.

Erzinclioglu, Zakaria. *Maggots, Murder, and Men.* New York: St. Martin's Press, 2002.

Genge, Ngaire. *The Forensic Casebook; The Science of Crime Scene Investigating.* New York: Ballantine Books, 2002.

Goff, Lee. *A Fly for the Prosecution.* Cambridge: Harvard University, 2000.

Hawthorne, Mike. *First Responder: A Guide to Physical Evidence Collection for Patrol Officers.* New York: CRC Press, 1998.

Hein, E.K. *Partners in Crime.* San Francisco: Jossey Bass, 2004.

Kleiner, Murray. *Handbook of Polygraph Testing.* New York: Academic Press, 2001.

Lee, Henry. *Henry Lee's Crime Scene Handbook.* North Carolina: Academic Press, 2001.

Manhein, Mary. *The Bone Lady: Life as a Forensic Anthropologist.* New York: Penguin Books, 2000.

Maples, William. *Dead Men Do Tell Tales.* New York: Broadway Books, 1994.

Moenssens, Andre, James Starrs, Carol Henderson, and Fred Inbau. *Scientific Evidence in Civil and Criminal Cases, Fourth Edition.* New York: The Foundation Press, 1995.

Randall, Brad. *Death Investigation: The Basics.* Tucson: Galen Press, 1997.

Sachs, Jessica Snyder. *Corpse: Nature, Forensics, and the Struggle to Pinpoint Time of Death.* Tennessee: Perseus Books Group, 2002.

Saferstein, Richard. *Criminalistics.* New Jersey: Prentice Hall, 2003.

Saukko, Pekka, and Bernard Knight. *Knight's Forensic Pathology, Third Edition.* London: Hodder Arnold Publications, 2004.

Spitz, Werner. *Spitz and Fisher's Medicolegal Investigation of Death: Guidelines for the Application of Pathology to Crime Investigation, Third Edition.* Springfield: C.C. Thomas, 1993.

Stuart, James. *Interpretation of Bloodstain Evidence, Second Edition.* New York: CRC Press, 2000.

———. *Scientific and Legal Applications of Bloodstain Pattern Interpretation.* New York: CRC Press, 1998.

Ubelaker, Doug. *Bones.* New York: M. Evans and Company, 1992.

Wonder, Anita. *Blood Dynamics.* Durham: Academic Press, 2001.

B

Web Sites

Bloodstain Analysis/Collection of Blood Evidence

www.bergen.org
www.bloodspatter.com
www.crime-scene-investigator.net

Chromatography

www.chromatography-online.org/2/contents.html
www.firearmsid.com/A_distanceGSR.htm
www.pharm.uky.edu/ASRG/HPLC/hplcmytry.html
www.shu.ac.uk/schools/sci/chem/tutorials/chrom/chrom1.htm
www.wcaslab.com/tech/tbicpms2.htm

DNA

www.craigmedical.com
www.cstl.nist.gov
www.dna.gov
www.ornl.gov
www.pcrlinks.com

Entomology

http://agspsrv34.agric.wa.gov
www.forensic-entomology.com
www.research.missouri.edu

Evidence Collection

www.crime-scene-investigator.net
www.fbi.gov
www.ojp.usdoj.gov

Fingerprints

http://bci.utah.gov
http://onin.com
www.forensic-evidence.com
www.policensw.com

Forensic Pathology

http://medstat.med.utah.edu
www.crimelibrary.com
www.deathsacre.com
www.forensiconline.com
www.pathmax.com

Gunshot Wounds

www.forensicmed.co.uk
www.surgical-tutor.org

Polygraph

http://truth.boisestate.edu
www.police-test.net
www.polygraph.org

Programs

www.aafs.org
www.allcriminaljusticeschools.com
www.msu.edu